Animal Energetics

TERTIARY LEVEL BIOLOGY

A series covering selected areas of biology at advanced undergraduate level. While designed specifically for course options at this level within Universities and Polytechnics, the series will be of great value to specialists and research workers in other fields who require a knowledge of the essentials of a subject.

Titles in the series:

Experimentation in Biology	Ridgman
Methods in Experimental Biology	Ralph
Visceral Muscle	Huddart and Hunt
Biological Membranes	Harrison and Lunt
Comparative Immunobiology	Manning and Turner
Water and Plants	Meidner and Sheriff
Biology of Nematodes	Croll and Matthews
An Introduction to Biological Rhythms	Saunders
Biology of Ageing	Lamb
Biology of Reproduction	Hogarth
An Introduction to Marine Science	Meadows and Campbell
Biology of Fresh Waters	Maitland
An Introduction to Developmental Biology	Ede
Physiology of Parasites	Chappell
Neurosecretion	Maddrell and Nordmann
Biology of Communication	Lewis and Gower
Population Genetics	Gale
Structure and Biochemistry of Cell Organelles	Reid
Developmental Microbiology	Peberdy
Genetics of Microbes	Bainbridge
Biological Functions of Carbohydrates	Candy
Endocrinology	Goldsworthy, Robinson and Mordue
The Estuarine Ecosystem	McLusky
Animal Osmoregulation	Rankin and Davenport
Molecular Enzymology	Wharton and Eisenthal
Environmental Microbiology	Grant and Long
The Genetic Basis of Development	Stewart and Hunt
Locomotion of Animals	Alexander

TERTIARY LEVEL BIOLOGY

Animal Energetics

ALAN E. BRAFIELD, B.Sc., Ph.D.
Senior Lecturer in Biology
Queen Elizabeth College
University of London

MICHAEL J. LLEWELLYN, B.Sc., Ph.D.
Lecturer in Biology
Queen Elizabeth College
University of London

Blackie

Glasgow and London

Distributed in the USA by
Chapman and Hall
New York

Blackie & Son Limited
Bishopbriggs, Glasgow G64 2NZ
Furnival House, 14–16 High Holborn, London WC1V 6BX

Distributed in the USA by
Chapman and Hall
in association with Methuen, Inc.,
733 Third Avenue,
New York, N.Y. 10017

British Library Cataloguing in Publication Data

Brafield, Alan E.
   Animal energetics.—(Tertiary level biology)
   1. Bioenergetics
   I. Title   II. Llewellyn, Michael J.
   III. Series
   591.19'.121   QH510

   ISBN 0-216-91255-5
   ISBN 0-216-91254-7 Pbk

Library of Congress Cataloging in Publication Data

Brafield, A. E.
   Animal energetics.

   (Tertiary level biology)
   Bibliography: p.
   Includes index.
   1. Bioenergetics.   2. Energy metabolism.
   I. Llewellyn, Michael J.   II. Title.   III. Series.
   QH510.B725   591.19'121   81-22516

   ISBN 0-412-00021-0 (Chapman and Hall)   AACR2
   ISBN 0-412-00031-8 (Chapman and Hall: pbk.)

Filmset by Advanced Filmsetters (Glasgow) Ltd

Printed in Great Britain by
Thomson Litho Ltd, East Kilbride, Scotland

Preface

All animals, from protozoans to primates, utilize energy in every aspect of their lives. Metabolic processes transform energy within the living cells. Individual animals obtain energy with their food and use it in a variety of ways. Populations and communities pass energy through their ecosystem. The growing interest in animal energetics is therefore seen in three major areas of zoology—biochemistry, physiology and ecology. We have tried to describe the main features of almost all these aspects, with the aim of arousing or feeding an interest in energetics in general as well as supplying some information on particular parts of the subject. The substantial list of references is intended to guide the reader to larger or more specialized works.

In covering such a wide field in so small a compass we will have been guilty of over-generalizations, but we find when teaching energetics, whether physiological or ecological, that students easily lose sight of the principles if all statements are hedged about with every relevant condition and exception. We have drawn examples from a wide range of animals, avoiding over-emphasis on mammals, but we have tended to favour those groups with which we are more familiar.

We are grateful to many people for helpful discussions and in particular to Kenneth Denbigh, Derek Miller and John Stirling for valuable criticisms of parts of the manuscript. We are responsible for all remaining errors, of course, and will welcome having them drawn to our notice.

A.E.B.
M.J.L.

Contents

CHAPTER ONE

ENERGY AND ENERGETICS

Energy is often defined as the capacity to do work, and work as the action
of a force in moving a mass through a distance. Neither definition is
very satisfactory and it is not easy to get the concept of energy clear in the
mind. This is partly because energy cannot be measured directly. We
can only measure the transformation of energy from one of its forms into
another.

The concepts of potential and kinetic energy are fairly familiar. Potential
energy is stored energy, possessed by a stick of dynamite, or a molecule of
glycogen, or a stone in a position from which it may fall. Kinetic energy is
possessed by a body by virtue of its motion. There are various forms of
energy—chemical, electrical, mechanical, radiant—and it is common
experience that one form can be converted into another. Energy in the
chemical bonds of a food substance may be converted into the tensile
energy of a working muscle and then be converted into thermal energy
when the muscle relaxes. Work and heat are not forms in which energy is
"held" (like, say, chemical energy) but are the means by which energy may
be passed between a system and its surroundings. In thermodynamics a
system comprises all the materials taking part in the process under
investigation or discussion. In biological terms a system might be an
organelle, a cell, an animal, or a whole ecosystem. An "open" system can
exchange both matter and energy with its surroundings, an "isolated"
system can exchange neither, and a "closed" system can exchange energy
but not matter.

Thermodynamics is the study of the interrelations of energy and matter.
The physicist is concerned with heat engines, expanding gases driving
pistons, electrical batteries and the like. For such matters the reader
should consult books on thermodynamics written by and for biologists
(e.g. Spanner, 1964; Ramsay, 1971). Here we need only concern ourselves
with a brief consideration of the laws of thermodynamics.

1.1   Entropy, free energy and enthalpy

Although living systems are bewilderingly complex and variable they nevertheless obey physical principles. Animals conform to the laws of thermodynamics as rigidly as do gases and machines. The first law, framed over a century ago, states that energy may be converted from one form to another but is neither created nor destroyed. Thus the total energy in a system and its surroundings is held to remain constant. One aspect of this principle is particularly important and useful to biologists and is known as the law of constant heat sums, or the law of Hess. This states that chemical reactions which start and end with the same substances liberate (or consume) the same amounts of heat, no matter which pathway is followed (so long as no changes in other forms of energy are involved). If the energy of the initial state exceeds that of the final one, the difference between the two must represent the energy lost as heat, a fixed amount regardless of the procedure by which the initial state became the final one. The complete oxidation of a mole of glucose to water and carbon dioxide, for example, liberates the same amount of heat whether this is done by fermentation via ethyl alcohol or by glycolysis and the Krebs cycle. This is convenient for biologists interested in energetics, because it means that if we know the difference in energy content between the initial and final states, the actual pathway followed is, in a sense, irrelevant.

The first law of thermodynamics says nothing of the extent or direction of energy transformations. The second, however, states that the disorder or randomness of the universe is continuously increasing, because during energy transformations some energy is degraded to a more random form, heat. Thus the inefficiency of energy conversions, whereby some energy is lost as heat, increases universal disorder, since heat is essentially the kinetic energy of random motion at the molecular level. This concept of randomness and disorder is known as *entropy*.

Entropy can be linked to another concept, that of *free energy*, by the following equation (applicable at constant temperature):

$$\Delta U = \Delta F + T\Delta S \tag{1.1}$$

where ΔU is the change in the total energy of a system, T the absolute temperature, and ΔS the change in entropy. ΔF represents the change in Helmholtz free energy, and represents the maximum work that the process involved is capable of doing (under isothermal conditions).

Another concept, which will recur later, is *enthalpy*. Change in enthalpy can be defined for our purposes as the amount of chemical energy in a system which is liberated as heat during combustion under constant pressure. Burning a substance in an open (constant pressure) calorimeter will yield a true value for the enthalpy change. In a constant volume bomb calorimeter (described in section 3.5) true enthalpy is not measured, because the pressure changes, but the error involved is small. Change in the enthalpy of a system (ΔH) is related to the change in entropy (ΔS), under conditions of constant temperature, by the equation

$$\Delta H = \Delta G + T\Delta S \tag{1.2}$$

where ΔG represents the change in Gibbs free energy (named after J. Willard Gibbs, a pioneer of chemical thermodynamics). Whereas ΔF, the change in Helmholtz free energy, is a measure of the *total* work possible (including any work involved in increasing the volume of a gas against atmospheric pressure) ΔG is a measure of the maximum *useful* work possible. Thus, at constant pressure

$$-\Delta G = -\Delta F - P\Delta V \tag{1.3}$$

where $-\Delta G$ represents the decrease in Gibbs free energy, $-\Delta F$ the fall in Helmholtz free energy, and $P\Delta V$ the work done by the system in increasing the volume (ΔV) against constant pressure (P). From equations 1.1 and 1.3 it can be seen that ΔF equals $\Delta U - T\Delta S$ (at constant temperature) and also equals $\Delta G - P\Delta V$ (at constant pressure). Thus

$$\Delta G = \Delta U - T\Delta S + P\Delta V \tag{1.4}$$

when both temperature and pressure are constant.

Biologists generally think in terms of ΔG (change in Gibbs free energy) rather than ΔF (change in Helmholtz free energy). It can be seen from equations 1.1 and 1.2 that ΔF is related to change in the total energy of the system (ΔU) in the same way as ΔG is related to change in enthalpy (ΔH). Nearly all energy changes in systems that biologists study take place under conditions of nearly constant temperature and pressure. In most metabolic processes ΔG and ΔF are almost the same, so the distinction between them need not concern us. Changes in free energy can be calculated for biochemical reactions, and so can be used to assess their direction and equilibria, but we will be mainly concerned with the energetic aspects of animal physiology and ecology rather than biochemistry, and so it will not be necessary to consider free energy very much further. The energetics of biochemical processes are well covered by Lehninger (1971), and Spanner (1964) gives full consideration to the difficult concepts of free energy and entropy.

1.2 Work and heat

The total energy (U), free energy (F or G) and entropy (S) of a system are all "properties" of the system. Work and heat are not properties in this sense. They are not forms in which energy is "held" but, as already pointed out, they are modes in which energy is passed between a system and its surroundings.

Conversions of "useful" energy from one form to another are never, in practice, completely efficient. Complying with the second law of thermodynamics, all such conversions involve degradation of some "useful" energy into heat. All work, including the activities and growth of animals, involves transformations of energy accompanied by production of heat. So heat is "inferior" to work. A system which can do work can transfer energy to any other system, but one which has only heat to offer can only transfer it to a system at a lower temperature.

Work and heat were first related quantitatively in 1845, when Joule connected a weight by cord and pulleys to a paddle immersed in water. In falling, the weight stirred the paddle and warmed the water, and Joule compared the work done by the falling weight with the rise in temperature of the water. In modern units, it turns out that 1 kg falling 1 metre (1 mkg of work) produces about 2.34 calories of heat in the water (enough to raise the temperature of $23.4 \, cm^3$ water by 0.1°C). The force (mass × acceleration) required to move 1 kg vertically through 1 metre is $1 \, kg \times 9.81 \, m \, sec^{-2}$, or 9.81 newtons. The work done (force × distance) is 9.81 newtons × 1 metre, or 9.81 joules. As about 2.34 calories are equivalent to 9.81 joules, 1 calorie equals about 4.19 joules. In terms of rate, 1 kcal (1000 calories) per hour is equivalent to about 1.16 joules per second, or 1.16 watts.

This raises the question of which unit is the most appropriate for use in animal energetics. The calorie has a long history as the unit for heat, but the joule is claimed by many to be the most suitable unit for general application in energetics. The position is complicated by the fact that there are several calories (for example the 15°C calorie and the mean calorie) and several joules (such as the international joule and the absolute joule), definitions of which can be found in standard texts. One 15°C calorie equals 4.1850 international (electrical) joules, or 4.1858 absolute (mechanical) joules; whereas the thermochemical calorie is equal to 4.184 joules exactly. In chemistry and physics measurements can usually be made very precisely, but in physiology and ecology the level of accuracy in measurement can be alarmingly low. There seems no point in worrying about the third or fourth decimal place in interconverting calories and joules, therefore, when precision elsewhere in the study may be rather poor. There is a good case for confining the calorie to use as the measure of heat and for using the joule as a sort of common denominator when adding together various forms of energy (Kleiber, 1972) as, for example, in the energy budgets described in the next chapter. In the generally accepted SI system of units the joule is the recommended unit for energy in all its manifestations, but it seems likely that the calorie will persist for some time as it is so familiar and has been so widely used. We use joules in this book, applying where necessary a conversion factor of 1 calorie to 4.184 joules (1 joule to 0.239 calories).

1.3 The study of energetics

Bioenergetics is concerned with energy transformations in living systems. It sheds light on the mechanisms by which all three major grades of organization operate: the cell, the organism and the population. Thus energetics plays a role in biochemistry, physiology and ecology.

The cells of animals carry out three basic kinds of work. First, there is the chemical work involved in biosynthesis, the building up of large and complex molecules (such as proteins and polysaccharides) from small and simple ones. Second, there is the work involved in transporting and concentrating materials. Cells take up certain substances against a concentration gradient, and eject others, and this process of active transport requires energy. Third, and most obvious, is the mechanical work done by contracting muscles (and by cilia and flagella). These three kinds of work represent energy transformations which basically fall within the realm of biochemistry. We will consider them only briefly (in chapter 4) because this book is primarily concerned with the uptake and loss of energy by animals, and populations of animals, and the utilization of energy resources in terms of animal or population growth and activity.

SUMMARY

1. Energy can be defined as the capacity to do work, and work as the action of a force when moving a mass through a distance. Potential energy is stored energy, for example that in the bonds of a chemical compound. Kinetic energy is energy of motion manifest in, say, a running animal.

2. Animals obey the laws of thermodynamics. The first of these states that energy may be converted from one form to another but is neither created nor destroyed. The second law concerns the "degradation" of some energy into heat during energy transformations. The

concepts of entropy ("disorder"), free energy (the work a process is capable of doing) and enthalpy ("heat content") are linked together by simple equations.

3. Work and heat are modes in which energy is passed between a system and its surroundings. (A cell or a whole animal, for example, can be a "system".) Work and heat are quantitatively related: 4.184 joules are equivalent to one calorie. The officially recommended unit for energy in all its manifestations is now the joule, but the calorie is likely to persist in some contexts as its use has been so widespread.

4. Bioenergetics is concerned with energy transformations in living systems and sheds light on the mechanisms by which all three grades of animal organization operate: the cell, the individual and the population. Animal energetics therefore plays a role in biochemistry, physiology and ecology.

CHAPTER TWO

THE ENERGY BUDGET

While reading this book you are expending energy. Eye muscles are active as the page is scanned and the print focused, and arm muscles are working as the pages are turned. In fact all the muscles are in tonus and so are not completely idle. Muscles of the gut are active as food is passed along it. Products of digestion are being absorbed by active transport in the intestine, and ions are being moved against diffusion gradients in, for example, nerve axons and the kidneys. Energy is being expended in maintaining the living cells, by breaking down some complex metabolites and building up others. In addition, in mammals, the body is being maintained at a higher temperature than the surroundings. When running or otherwise "working hard" the muscles are far more active, of course, and the rate of heat production is consequently much higher. A great deal of an animal's energy expenditure occurs in the muscles. Muscular activity is mainly concerned with locomotion, as the animal moves about its habitat seeking food or a mate, or avoiding predators. Thus, as a result of their movements, animals generally expend far more energy than do plants.

Energy is derived from the food an animal eats. So food provides an animal with *two* vital commodities—chemicals and energy. We will call the energy of the food C, standing for consumption. If the animal is growing, some energy will be retained in the body in the chemical bonds of the growth materials (called P, for production). Some energy will presumably be passed to the environment as heat (Q) and through work (W) done by the animal on its surroundings (section 1.2). In addition, energy will be lost in the chemical bonds of waste products in the urine (U) and in the faeces (F). As energy can neither be created nor destroyed, these considerations suggest the following schematic equation:

$$C = P + Q + W + U + F \qquad (2.1)$$

7

The energy lost as heat is usually called R rather than Q, as it is usually assessed by measuring the rate of respiration (chapter 5). Work done on the surroundings is generally negligible, for reasons given in the next section, and so W can nearly always be safely ignored. Thus we can write

$$C = P + R + U + F \qquad (2.2)$$

This is the equation first framed by Petrusewicz and Macfadyen (1970) and now generally used by physiologists and ecologists to represent the energy budget of an animal or a population.

2.1 The thermodynamic basis of the energy budget

Equation 2.2, expressing the factors in an energy budget, is based on biological reasoning which might be imprecise or incomplete. Can it be derived from a rigorous application of the laws of thermodynamics, which are known to govern energy changes and transfers in the non-living world? According to the first law (section 1.1) energy exists in various forms but cannot be created or destroyed. Energy may be transferred in either of two modes—heat and work. Thus in a "closed" thermodynamic system, exchanging energy but not matter with the surroundings,

$$\Delta U = Q + W \qquad (2.3)$$

where ΔU is the change in energy content of the system, Q the net exchange of heat with the surroundings and W the net exchange of work with the surroundings. For an "open" system, exchanging both energy and matter with the surroundings (as in a feeding animal), one can write

$$\Delta H_s = \Delta H_m + Q + W \qquad (2.4)$$

where ΔH_s is the overall change in enthalpy of the system (animal) and ΔH_m is the net exchange with the surroundings (environment) of enthalpy of matter. This concept is shown in diagrammatic form in figure 2.1, from which it can be seen that equation 2.4 can be written

$$\Delta H_s = (H_1 - H_2) + (Q_1 - Q_2) + (W_1 - W_2) \qquad (2.5)$$

There will be dissipation of internal work, such as that involved in biosynthesis, active transport and so on, as heat. This will eventually be passed on to the surroundings, and so form part of Q_2. Work done on the environment, W_2, is almost always negligible. Termites building high nests of consolidated soil particles, and beavers raising dams, are both cases of work being done on the surroundings, as the animals are increasing the

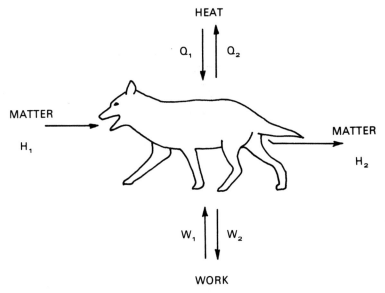

Figure 2.1 Diagram of the concept of an animal as an open thermodynamic system exchanging energy (as heat and work) and matter with its surroundings.

potential energy of the materials they raise. Such examples are rarities, however, and even in these cases the work done on the environment is only a tiny fraction of the total work expended by such animals in other ways. An exception is a horse pulling a load. Here work is being done by the animal in overcoming friction. This energy will be manifest as heat in the surroundings. Work will also be done if the horse is going uphill, for it will then be increasing the potential energy of the load. (Brody, 1945, discusses the work done by draught animals.) In the great majority of cases, however, W_2 is so small that it can safely be ignored. Work done by the environment on the animal (W_1), such as an upwelling ocean current raising the position of a member of the plankton, or an updraught increasing the height of an insect or a bird, is small and of rare occurrence (particularly as the reverse process is likely to occur sooner or later, restoring the animals to their original position). Thus the factor W_1 can also be disregarded. Equation 2.5 can then be written

$$\Delta H_s = H_1 - H_2 + Q_1 - Q_2 \tag{2.6}$$

Now ΔH_s, if positive, is equivalent to the factor P of equation 2.2, the energy laid down in the growth material of a feeding and growing animal.

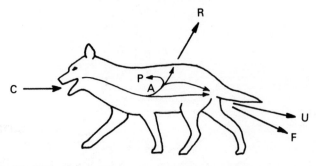

Figure 2.2   Idealized concept of the energy budget for an animal. Energy in the food (C) is retained in growth material (P) or lost in the faeces (F), in urine (U) and as heat (R). A is assimilated energy (i.e. C − F). It is assumed that work done by the animal on its surroundings, and vice versa, is negligible.

H_1 is the energy content of the food eaten, or C in equation 2.2. H_2 is the energy in matter leaving the animal, which will be contained in the urine (U) and the faeces (F). Furthermore, the net exchange of heat between the body and its surroundings, $Q_1 - Q_2$, will be negative because the heat lost (Q_2) will normally exceed that gained (Q_1). This is equivalent to R in equation 2.2, the net heat loss, and so one may substitute in equation 2.6 and write

$$P = C - (U + F) + (-R) \qquad (2.7)$$

which when rearranged gives

$$C = P + R + U + F \qquad (2.2)$$

Thus the equation for the energy budget compiled earlier, on biological grounds, is established on a sound thermodynamic basis. This derivation has been fully and lucidly discussed by Wiegert (1968). The factors in the equation for the energy budget are shown diagrammatically in figure 2.2.

2.2   Energy budget of an individual

Figure 2.3 shows an energy budget for a perch (*Perca fluviatilis*) over a 28-day period, based on the results obtained by Solomon and Brafield (1972) from one of a number of similar experiments. Food was in the form of the small crustacean *Gammarus*. The energy in the food eaten over this period (C), the energy retained as growth (P) and the energy content of all the faeces produced (F) were calculated by means of bomb calorimetry

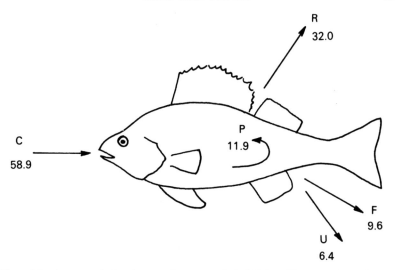

Figure 2.3 An energy budget for a perch over a period of 28 days. Figures are in kilojoules.

(see section 3.5). Energy lost in the urine (U) was estimated by measuring the amounts of ammonia excreted and multiplying by the known energy content of ammonia. Respiratory heat loss (R) was obtained by recording the oxygen consumption continuously over the 28 days. The total amount of oxygen consumed was then multiplied by an oxycalorific coefficient to arrive at an estimate of the amount of energy lost as heat. This technique is known as indirect calorimetry, and is discussed in section 5.1. (In some circumstances it is possible to measure the heat loss directly, by measuring the rise in temperature of the surroundings. This process is also described and assessed in section 5.1).

It will be noticed that the sum of the values for P, R, U and F in figure 2.3 is 59.9 kilojoules (kJ), whereas the energy of the food was estimated to be 58.9 kJ. These two figures should, of course, be identical, as the equation for an energy budget states that $C = P + R + U + F$. Clearly some error arose in the measurement of one or more of the channels. The balance of this budget is satisfactory, however, when one considers the variety of techniques employed and the potential sources of error involved. The extent of budget balance is often expressed as a percentage, and the results of figure 2.3 yield a ratio of 59.9 to 58.9 kJ, or 101.7 %.

The erection of such a budget for a feeding and growing animal serves to indicate, among other things, the proportion of energy at the animal's disposal which is retained as growth (P) and that which is used in what one may broadly term activity (R). In the case of a starved animal, food energy (C) and faecal energy (F) will be zero and growth (P) will therefore be negative, because the animal will lose weight (release stored energy) in order to provide energy for R. It follows that there is a certain level of feeding at which an animal will neither gain nor lose weight (P is zero). This level is known as the *maintenance ration* (section 3.3).

2.3 Energy budget of a population

A population of animals, that is to say an interacting group of individuals of the same species, represents an open thermodynamic system in much the same way as an individual animal does. Thus equation 2.2, embodying the concept of the energy budget, should apply to a population as well as to an individual. A population feeds and grows and loses energy in faeces and excreta and as heat, just as an individual does. Sometimes all these channels can be estimated, and a population energy budget erected, but the difficulties are far greater than in the case of an individual. Information is needed on population size, age structure, birth rate, death rate and the like, and this is often difficult to obtain. Consequently the erection of an energy budget for a population is subject to greater uncertainty than in the case of an individual (see section 8.6). Where this can be done fairly reliably, however, valuable insight is gained into how the available energy is allocated among the various channels, just as with an individual energy budget. Growth and activity can be compared, for example, in that these channels are represented by P and R. A very active species is likely to be rather slow-growing, as it is "spending its income rather than banking it", whereas a sluggish one may be able to grow more rapidly. Thus, as a crude simplification, P and R can be seen as competing with one another for the resource C.

An example of a population energy budget is shown in figure 2.4. It reflects estimates made by Kay and Brafield (1973) for a population of the polychaete *Neanthes virens* in an estuarine mudflat. Balance is exact (i.e. C exactly equals $P + R + U + F$) because all channels were not independently estimated and so some had to be derived from others. Another example of an energy budget for a population is shown in figure 2.5, reflecting a study by Llewellyn (1972) of the energetics of a population of the aphid *Eucallipterus tiliae* on a lime tree. It will be seen that the energy lost in faeces (F) could not be separated from that of the nitrogenous waste products (U) and this is commonly the case where, as in insects, the latter

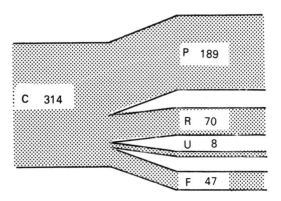

Figure 2.4 An energy budget for a population of a polychaete worm (*Neanthes virens*) in an intertidal mudflat. The figures are estimates of energy flow expressed as kJ m^{-2} year^{-1}.

are passed into the gut, where they mix with the faecal material. On the other hand, the energy retained in growth (P) can often be subdivided into a number of distinct categories. In this case, Llewellyn was able to distinguish body growth (Pg), reproductive production (Pr) and exuvial (moulted cuticle) production (Pe), which occurred in a ratio of about 6.5 : 1.0 : 1.2. The three are grouped together as P in figure 2.5.

The energy actually assimilated by an animal or a population, A, is the difference between the energy in the food (C) and that in the faeces (F). Expressing this as a percentage of C gives the *assimilation efficiency*. Thus

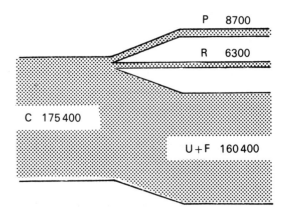

Figure 2.5 An energy budget for a population of an aphid (*Eucallipterus tiliae*) on a lime tree. Figures are in kJ year^{-1}.

in the case of the perch (figure 2.3), the assimilation efficiency equals $(58.9 - 9.6)/58.9$, or 83.7%. That for the *N. virens* population (figure 2.4) is $(314 - 47)/314$, or 85.0%. The assimilation efficiency of the aphid population (figure 2.5) cannot be accurately calculated, as F is not distinguished from U, but the nitrogenous waste is likely to be very small compared with the loss of energy in the faeces, and so the assimilation efficiency can be assumed to be extremely low, as F has almost as high a value as C. This is because large quantities of sucrose from the phloem sap, on which the aphids feed, pass straight through the gut and leave the insects as honeydew. Much sugar is therefore "wasted" by the insects in their "search" for valuable but scarce amino acids.

2.4 Energy flow in the ecosystem

Populations of the various plant and animal species in a particular area constitute the local community. The community and the region it occupies affect one another, and together they form the ecosystem. The energetics of an ecosystem, such as a coniferous forest or a coral reef, can be seen in terms of one enormous energy budget. As figure 2.6 indicates, a small amount of the solar energy which enters the ecosystem is trapped by plants in the process of photosynthesis, providing for the production (growth) of the plant species. Some of this energy will be taken in as food by herbivorous animals, and a proportion of the energy in this food will become locked in the tissues of the growing animal. Carnivores, in their turn, will utilize some of this energy after feeding on the herbivores. Thus energy passes from one trophic level to another, but a great deal is lost as heat during each transformation. This heat cannot be utilized by the living components of the ecosystem, and so although materials (elements and chemicals) can be recycled within an ecosystem, there is essentially a one-way flow of energy. The energetics of ecosystems is a highly complex field of study, and ecosystems differ widely in the form and scale of the various energy pathways involved.

2.5 The energy budget as a basis for the study of animal energetics

The concept of the energy budget provides the central theme of this book. The attraction of the concept lies in its essential simplicity, as expressed in equation 2.2; its usefulness lies in the fact that it can be applied in principle to *any* animal or population. Framed and determined by the laws of thermodynamics, the energy budget forms a firm basis for investigation

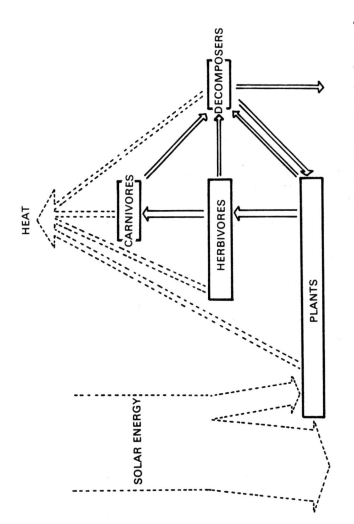

Figure 2.6 Simplified flow diagram of the energetics of an ecosystem. Materials can recirculate by way of decomposers. Energy flow, on the other hand, is essentially one-way, as energy is ultimately degraded and lost as heat. Solid arrows indicate transfer of both materials and energy, dotted ones energy alone.

and experimentation. It affords a method of obtaining insight into the various ways in which energy is utilized by animals, illuminating both the nature and the relative sizes of these channels. Each of the factors in the equation for the energy budget has its own peculiar features of interest (and difficulties in measurement and assessment), and each is considered in turn in the following chapters, before being brought together again in the closing ones.

SUMMARY

1. A feeding animal is an open thermodynamic system, exchanging both matter and energy with its surroundings. The overall change in enthalpy of the animal is equal to the sum of net exchanges with the environment through three agencies: heat, work, and the enthalpy of matter. Since work done by the animal on its environment and work done by the environment on the animal are generally negligible, it can be shown that for nearly all practical purposes $C = P + R + U + F$, where C (consumption) is the energy content of the food eaten, P (production) the energy content of growth material, R (respiration) the energy lost as heat, U (urine) the energy content of excreted nitrogenous waste and F the energy content of the faeces. This equation represents the energy budget of the animal.

2. An energy budget for an animal can be compiled by calculating each of the five factors in the budget equation. It contains much information, such as the proportion of energy at the animal's disposal which was retained as growth.

3. A population of animals also represents an open thermodynamic system, and if enough data can be collected a population energy budget can be erected. Such budgets yield interesting information about the ways in which different species allocate their energy resources among the various pathways at their disposal.

4. The populations of various animals and plants in a particular area constitute the local community. The community and its environment affect one another, and together they form an ecosystem. The energetics of an ecosystem can be seen in terms of one enormous energy budget. Energy from solar radiation is trapped by plants and passes through herbivores and then carnivores. Much of it is ultimately returned to the environment as heat.

5. The concept of the energy budget, derived from the laws of thermodynamics, forms a firm basis for many and varied studies in animal energetics. It provides insight into the various ways in which energy is utilized and the relative sizes of these channels.

CHAPTER THREE

ENERGY INTAKE

The food an animal eats provides it with two essential ingredients: the elements and compounds needed to maintain and increase its structure, and the energy required to drive its metabolic machinery and operate its muscles. The energy content of the food must be sufficient to provide for the energy requirements of the animal. Energy intake, represented by C (for consumption) in the equation for the energy budget presented in the previous chapter, is the subject of this one.

3.1 The energy available to animals

A man requires about 9800 kJ per day from his food, and 12 500 kJ is a generous allowance. The energy is chiefly supplied by the carbohydrates, fats and proteins in the diet, after these have been digested and absorbed. In a "well-balanced" human diet 50–70% of the energy is in the form of carbohydrate, and the energy needs of the living cells are generally met by the respiration of carbohydrates. Energy can also be released through the respiration of amino acids and fats, however.

Biologists usually express the energy content of a food material as the *heat of combustion*. This is the heat produced when the material is completely oxidized in a bomb calorimeter (section 3.5). The products of such combustion are carbon dioxide and water and, in the case of protein, nitrogen. The heats of combustion of various substances are shown in table 3.1. It can be seen that fat has more than twice the energy value of carbohydrate and that protein lies between the two. This is because of the different ratios of carbon to oxygen. Oxygen provides about 10% by weight of a fat (e.g. 10.9% for oleic acid) but about 50% of a carbohydrate (e.g. 53.3% for glucose). As the carbon to oxygen ratio of a fat is relatively high, to effect combustion relatively more oxygen must be supplied from

Table 3.1 Some heats of combustion

	kJ g^{-1}	kJ mole^{-1}
Protein (mean)	23.6	—
amino acids:		
alanine	18.2	1619
glycine	13.0	975
tyrosine	24.8	4485
Fat (mean)	39.5	—
fatty acids:		
oleic	39.7	11 209
palmitic	39.1	10 017
stearic	39.9	11 326
Carbohydrate (mean)	17.2	—
glycogen	17.5	—
starch	17.7	—
sucrose	16.6	5665
glucose	15.7	2833

outside—from high oxygen pressure in the bomb calorimeter—and more heat is produced in the process.

In plant seeds and similar storage organs one or two materials often predominate, and their energy values reflect this. Maize meal is high in carbohydrate and has a value of 18.54 kJ g^{-1} whereas soybean, which is chiefly protein and fat, contains 23.1 kJ g^{-1}. Whole animals and plants generally show less variation in their energy content, for they tend to contain broadly similar mixtures of protein, carbohydrate and fat. If the material left unburned in a bomb calorimeter is weighed, energy values can be expressed in terms of ash-free dry weight. Most plants show values between 17.78 and 21.71 kJ g^{-1} ash-free dry weight, with a mean of 19.75. Values for whole animals, taking the full range from protozoans to mammals into account, mainly fall between 21.38 and 26.15 kJ g^{-1} ash-free dry weight, with mean of 23.77 (d'Oleire-Oltmanns, 1977). Thus most carnivores are tapping a more "energy-rich" food source than those animals which eat plants, assuming the proportion of incombustible material is the same in each case. Energy values in terms of ash-free dry weight are larger than those expressed in terms of total dry weight, by an amount reflecting the quantity of ash in the sample. Mean values of 17.87 and 20.54 kJ g^{-1} dry weight have been calculated for plant and animal material respectively (d'Oleire-Oltmanns, 1977).

The energy content of the food is not all available to the animal as useful energy, as can be seen from figure 3.1. Incomplete digestion and absorption means that some energy is lost with the faeces. If the energy contents of food and faeces are known, the percentage of food energy which is absorbed in the gut can be calculated. This is the assimilation efficiency, which nutritionists tend to call the digestibility coefficient. Assimilation

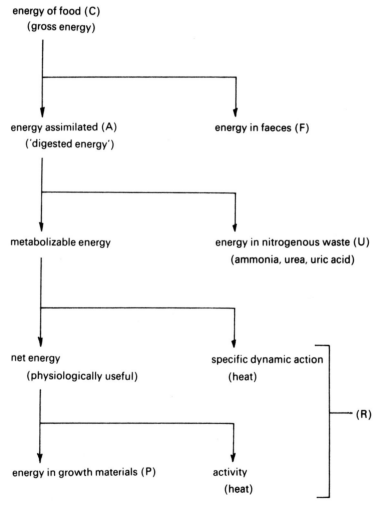

Figure 3.1 Not all energy intake (energy in the food eaten) is of use to the animal. The net energy in the above scheme is likely to be only about 60% of the gross energy in the case of protein. Letters in brackets are factors in the energy budget as shown in equation 2.2. It is assumed that no external work is being done (see chapter 2).

efficiencies vary widely according to the nature of the food eaten by the animal (see table 8.2), carnivores generally showing higher values than herbivores.

Many of the simple products of digestion which are absorbed by the gut are sooner or later broken down. The waste products of carbohydrate and fat catabolism are carbon dioxide and water, but in the case of protein some nitrogenous waste is produced as well, and this contains some energy. The commonest such waste products are ammonia, urea and uric acid, which have energy values of about 20.5, 10.5 and $11.5 \, \text{kJ} \, \text{g}^{-1}$ respectively. Of the 23.6 kJ resulting from combustion of a gram of protein (table 3.1), nearly 3.8 will be lost as urea in the case of a mammal, thereby reducing the useful energy to less than $20 \, \text{kJ} \, \text{g}^{-1}$.

Some of the metabolizable energy (figure 3.1) becomes lost as heat through a process commonly known as specific dynamic action (section 3.4). The physiologically useful energy, which is that remaining, can be called the net energy. This is significantly less than the gross energy, the energy of the food.

Intake of energy is not necessarily by way of the mouth and gut. Examples of animals which have no gut are intestinal parasites such as the tapeworms (cestodes) and acanthocephalans, and the pogonophorans (a group of deep-sea worms). In such cases nutrients must be taken in through the skin, and for this to occur they must be small molecules— amino acids rather than proteins, monosaccharides rather than poly-saccharides. Some animals which do have guts, however, also take in useful materials through the skin. It is becoming clear that many aquatic invertebrates living in habitats rich in dissolved organic matter can take in appreciable amounts of nutrients over the general body surface, often against considerable concentration gradients. Energy-requiring active uptake processes are involved and there seem to be separate mechanisms for glucose and for amino acid transport. The latter may be sodium-dependent. Active uptake of fatty acids has also been demonstrated. A wide range of animals has been investigated, but most work has concerned bivalve molluscs, polychaetes and echinoderms. The subject has been reviewed by Conover (1978) and Stewart (1979). There is some controversy as to whether the laboratory experiments are realistic, however, and it is uncertain whether animals in their natural habitats actually do, in fact, take up significant quantities of dissolved nutrients through the skin, although they have the ability to do so. The importance of this channel of energy intake cannot be established until net uptake of dissolved material is clearly demonstrated under field conditions.

Before considering how animals may regulate their energy intake, some points must be made about food availability and food choice. The availability of food in the animal's habitat is subject to many factors outside its control. The duration and intensity of sunlight will affect photosynthetic rates, for example, and so affect the production of plant food for herbivores; and factors which affect herbivore production and availability will affect the feeding of the carnivores which prey upon the herbivores. If food is in excess, an animal may choose the amount it eats, when it eats and what it eats. There may be a shortage of food, however, making it difficult to find, or causing competition for it. Some animals feed sporadically, particularly predators. Many herbivores tend to feed almost continuously, however, and so do filter-feeders (e.g. some polychaetes, bivalve molluscs and crustaceans). Finally there is the question of food quality. Animals that tap a very specialized food source (e.g. nectar, plant sap, blood) may have to eat excessive amounts of one commodity in order to obtain enough of another: aphids feeding on plant sap have to consume large quantities of sucrose to obtain enough amino acids, which are relatively scarce.

3.2 Regulation of energy intake

The factors that control the sensations of hunger or satiety that we experience are subtle and far from fully understood. Most experiments in this field have concerned mammals, chiefly rats and dogs, and it is clear that the hypothalamus plays a role. This region of the brain has a hunger centre or feeding centre in the lateral nucleus and a satiety centre in the ventromedial nucleus. Destruction of the former, in lesion experiments, can cause irreversible loss of appetite, starvation and death; whereas destruction of the latter induces overeating and obesity. Conversely, electrical stimulation of the lateral centre causes increased food intake, whereas stimulation of the satiety centre abolishes the hunger drive. Clearly these areas (and also, it seems, parts of the cortex and the medulla) play a part in regulating food intake, but what are the stimuli to which these cerebral centres respond? There is still no comprehensive answer to this question but various theories have been proposed to explain how energy intake is controlled. These have been discussed by Blaxter (1967) and Kleiber (1975).

The *glucostatic theory*, pioneered by Mayer, suggests that hypothalamic receptors monitor the glucose level of the blood (or perhaps the difference between the arterial and venous glucose levels), a fall in the blood glucose concentration stimulating the feeding centre and a rise stimulating the satiety centre. Increased electrical activity in the satiety centre, and diminished activity in the feeding centre, have been monitored after artificially raising the blood glucose level. Reducing the glucose level had the opposite effects. These artificial changes in blood glucose level were substantial (50% or more), however, and it is not certain that the very small changes that normally occur are sufficient to activate the hypothalamic centres. (Evidence is stronger that falling blood glucose levels stimulate the hypothalamic glucoreceptors and thereby cause mobilization of liver glycogen by the action of adrenalin.) There are some difficulties with the glucostatic theory, and perhaps the blood levels of metabolites other than glucose are involved. The *lipostatic theory*, proposed by Kennedy, suggests that the hypothalamus controls food intake in response to changes in body fat, receptors responding to circulating metabolites (perhaps fatty acids) which are in concentrations proportional to the extent of fat stores. Alternatively, metabolic waste products may be involved, for reduced food intake has been induced in ruminants (cattle) by perfusing ammonia into the blood.

In contrast to these chemostatic theories, Brobeck has proposed a *thermostatic theory*. This holds that eating is a response to a fall in heat production, whereas a rise in heat production causes the animal to stop feeding. There are undoubtedly thermoreceptors in the mammalian hypothalamus, concerned with regulation of body temperature (section 5.4). These receptors could be involved in influencing eating, although they are located in different areas to the feeding and satiety centres.

The chemostatic and thermostatic theories are relatively recent, mainly attracting attention in the fifties. An older idea suggested that contractions of an empty stomach signal hunger and initiate feeding, whereas distension of a full stomach causes cessation of feeding. Perhaps in some species the condition of the stomach does provide some coarse control of appetite, but in general the theory has been discarded, largely as a result of the experiments of Adolph, reported in 1947. He mixed the diet he gave to rats with various proportions of energetically inert materials such as kaolin, and found that the animals adjusted the intake of bulk in such a way that the energy intake was kept constant. So if there were any stomach distension signals, they were being over-ridden by a factor which controlled the level of intake of energy—the rats were "eating for calories". In addition, cutting afferent nerves from the stomach seems to have little effect on hunger, and so the condition of the stomach has fallen out of favour as a candidate for the source of control of food intake. Nevertheless, there is obviously a limit to the amount of food an animal can eat, as its alimentary canal can only hold so much at a time. Whereas an empty stomach has a small or uncertain influence on when an animal starts to feed, a full one must have some influence on when it stops. There are also psychological and social influences, in the case of man. In affluent societies, where food is plentiful and varied, palatability is an important factor, as is the wish to lose weight by reducing food intake.

Apart from mammals, most studies of the control of food intake have concerned insects, chiefly locusts and blowflies (Bernays and Simpson, 1981). Phagostimulants, commonly sugars, often initiate and maintain feeding. Insects stop feeding as a result of gut distension, sensory fatigue, and a decline in what entomologists call the central excitatory state. The neural and hormonal mechanisms which control feeding by insects are still imperfectly understood, however, as they are in the case of mammals.

3.3 Maintenance rations

It is evident from the energy budget equation $C = P + R + U + F$ (equation 2.2), that if energy intake (C) exceeds the energy lost from the body $(R + U + F)$ then the total energy held in the animal's body tissues must rise, and P will be positive. Similarly, if energy intake is less than energy loss, the energy content of the body will fall, and P will be negative. A rate of food energy intake which exactly balances energy losses is known as the *maintenance ration*. In such a case body size will tend to remain constant — body weight neither increasing (growth) nor decreasing. We say "tend to", because the average energy content per gram of an animal's tissues can vary. If body fat is increased at the expense of body carbohydrate, for example, the total body weight could remain the same although the total energy content has risen. In spite of the fact that the energy value per unit weight of body tissue is variable, it is nevertheless customary, because it is convenient, to think of growth in terms of increase in body weight. In intensive stock rearing and fish farming energy intake is manipulated, as far as this is economically feasible, so as to maximize increase in body weight.

Figure 3.2 shows two examples of experiments in which the energy of growth materials (or production, P) is compared with energy intake (consumption, C). The value for C which produces no increase or decrease in the energy content of the body (P is zero) is the maintenance ration. Rates of energy intake above and below this are known as super-maintenance and sub-maintenance levels respectively. In the first case P is

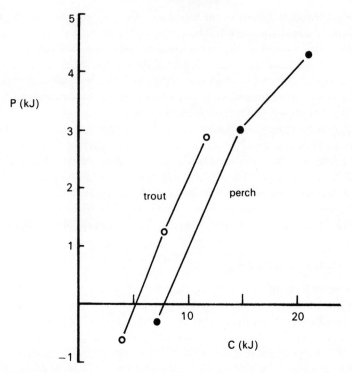

Figure 3.2 Two examples of experiments with fish which compare growth (P) with food intake (C). Maintenance rations (under the conditions described below) are indicated by the values for C when P is zero. Brown trout (*Salmo trutta*) rates per day, at 15°C, for about 50 g wet wt fish (calculated from Elliott, 1976). Perch (*Perca fluviatilis*) rates per 10 days, at 14°C, for 10–12 g wet wt fish (calculated from Solomon and Brafield, 1972).

positive, in the second negative. When C is zero, that is to say when the animal is starved, F (faecal energy) will also be zero, and so the negative value of P will be equal to the only two remaining factors in the energy budget equation—R and U. This is simply saying that in a starving animal all the energy lost, whether as heat (R) or in excretory products (U), has been provided by material already within the body.

The maintenance ration is a useful concept in laboratory experiments on energy balance, and also in a scientific approach to rearing fish or cattle or fowl. For an animal (or a population) in its natural environment, however, the maintenance ration may be impossible to assess, and the concept is of limited value to the ecologist in any case. The food available to the animal or population may be inadequate, for the reasons outlined in section 3.1.

3.4 Specific dynamic action

The metabolic rate of an animal is generally found to rise for a time soon after feeding. This increase can be seen by measuring either oxygen consumption or heat production, and there can also be a rise in the rate of nitrogenous excretion. This increase in metabolic rate after feeding has been called specific dynamic action (SDA), heat of nutrient metabolism, calorigenic effect of food and post-prandial thermogenesis. Nutritionists now tend to use the term dietary-induced thermogenesis (DIT).

The processes of feeding and digestion involve increased energy expenditure in manipulating the food and passing it along the gut, and in the secretion of digestive enzymes and the active uptake of the products of digestion, but specific dynamic action is usually on a scale far in excess of these demands. Rubner showed early this century that SDA is primarily due to metabolic processes rather than the work involved in feeding and digestion, but the exact cause of SDA is still debated. (Incidentally, Rubner called the process "specific dynamic effect" (*Wirkung*), which became mistranslated into English as specific dynamic action. The latter term has persisted, although specific dynamic effect should really be used.)

Specific dynamic action has been extensively studied in man, where metabolic rate may increase by about 30% after a meal and may stay elevated for an hour or two or much longer. Other mammals, particularly dogs, have been studied a great deal, and SDA is also characteristic of other vertebrates, notably fish. A good example has been described by Jobling and Davies (1980) for plaice (figure 3.3). They found that oxygen consumption rose after a meal to a maximum of about twice the resting rate, returning to the normal level after 24 to 72 hours. The extent and duration of the SDA increased with increasing proportion of protein in the diet. Even invertebrates have been found to show SDA. For example Bayne and Scullard (1977) kept mussels (*Mytilus edulis*) without food and then provided them with a suspension of unicellular algae for an hour. Part of the resultant rise in oxygen consumption was due to an increased filtration rate, but the rest was shown to be SDA.

Lavoisier was the first to measure an increase in oxygen consumption following a meal. The respiratory rate of his human subject was raised by about 50% over the fasting level. Later workers who repeated Lavoisier's experiment thought the effect was due to the work associated with

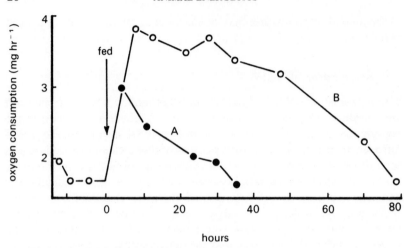

hours

Figure 3.3 An example of specific dynamic action (SDA). Oxygen consumption by plaice (*Pleuronectes platessa*) after being fed meals of fish paste with energy contents of 1.41 kJ (A) and 5.65 kJ (B). (After Jobling and Davies, 1980.)

digestion, but this was discounted early this century when it was found that intravenous injection of amino acids markedly increased metabolic rate. In 1888 Voit had suggested that increased circulation of nutrients in the blood triggered increased cell metabolism. This idea that high concentrations of metabolites were the cause of high rates of cellular activity (the plethora theory) was revived some years later by Lusk. Rubner, who has already been mentioned, held that the rise in metabolic rate was due to such processes as the deamination of amino acids in the liver. Working with dogs, he found that heat production was greater on days when meat was fed than when the diet was of fat or sugar. A broadly similar result was obtained by Lusk. The validity of experiments involving single-component diets has been questioned, however, and Garrow (1973) found very similar percentage increases in heat production in human subjects fed meals (of equal energy content) of one material only (sugar or fat or gelatin) or a mixture (protein and sugar). Garrow concludes that SDA is not specific for a particular nutrient, and suggests that it is related to the linking together of amino acids in protein synthesis rather than to their destruction by deamination. The theory that protein synthesis is responsible for SDA is a recent one (Ashworth, 1969), and receives support from studies with malnourished children, in which a high correlation has been found between the size of SDA and the rate of growth.

The picture is still unclear. SDA can often be matched with increased production of ammonia (by fish) or urea (by mammals), and so a link between SDA and the removal of "excess" amino acids cannot be ruled out. The idea that SDA reflects amino acid degradation (and urea synthesis, in the case of mammals) has a long history, Krebs being its most recent champion. SDA often occurs so soon after a meal that it cannot reflect metabolism of that meal's digested products and so must, at least in some cases, concern materials already present in the metabolic pool. Indeed SDA can occur at a time when a meal is normally given but is in fact withheld—a sort of conditioned or anticipatory SDA.

The literature on specific dynamic action is large, confusing, and sometimes contradictory. That no single theory is yet universally accepted is due partly to widely varied experimental techniques producing widely varied results, and partly to the variety of ways in which SDA is quantitatively expressed—as a proportion of the maintenance level, or of the fasting metabolic rate, or of the energy content of the meal. In addition SDA is so widespread a process that it is unlikely always to have exactly the same nature and cause. The height and duration of SDA can vary with different foods (fat, carbohydrate, protein), with the frequency of feeding and the size of meal, with the general level of activity, with temperature and species and age. Amino acids certainly appear to be involved, though whether it is their deamination and elimination or their use in protein synthesis (or both) which is responsible is still uncertain.

3.5 Bomb calorimetry

The energy values of some common food components and of plant and animal material in general were given in section 3.1, and we will now describe how such values are obtained. Accurate measurement of the energy content (heat of combustion) of biological material is vital in studies of energy balance. Energy intake can only be calculated when the energy value of the food is known, and when compiling an energy budget the energy contents of the animal and of the faeces are required. Chemical methods of measuring the energy content of biological samples, involving oxidation by strong acids, have been tried but are not very satisfactory. Instead the heat of combustion of a sample is usually measured with a bomb calorimeter.

Many kinds of bomb calorimeter have been built but the principle involved is the same in all. A sample of the material is dried, weighed, placed in a thick-walled steel chamber and completely burned by igniting it in oxygen at a high pressure. The heat released by the total combustion of the sample is calculated either by measuring the resultant increase in temperature of a surrounding water jacket or by means of thermocouples in close contact with the bomb's outer surface.

The bomb calorimeter shown in figure 3.4 is a miniature one similar to that designed by Phillipson (1964) for use when only small samples (as

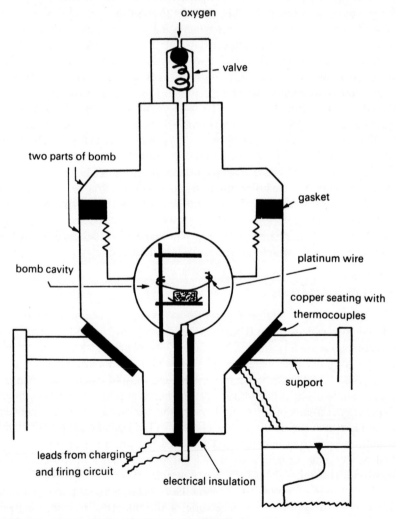

Figure 3.4 Diagrammatic view of a bomb calorimeter.

small as 10 mg) are available. (Most bomb calorimeters require about 200 mg samples.) Its use will be described in some detail as an example of the general technique. The material to be bombed is homogenized and dried, usually at 60–70°C. Lower temperatures fail to remove enough water and higher ones can cause the loss of volatile organic materials such as fats. The sample is then compressed into a pellet, weighed, placed on a small pan of platinum or nickel and put on the sample holder in the steel bomb. A length of fine wire, usually platinum, is fitted as shown in figure 3.4. The wire touches the sample pellet (or in some calorimeters connects to it by a small piece of cotton thread). The two parts of the bomb are tightly screwed together. The cavity is filled with oxygen to a pressure of 25 to 30 atmospheres and then encased in thermal insulation material such as preformed sections of expanded polystyrene. A firing circuit is used to put a sudden and large charge through the wire. This ignites the pellet and the heat of the combustion is taken up by the bomb casing. The rise in temperature is sensed by an array of thermocouples in the bomb's seating and the resulting voltage can be measured with a galvanometer or strip-chart recorder. The maximum deflection of the recorder pen (figure 3.4), in millivolts, is noted. Any unburned material (ash) is weighed. By burning a range of differently sized pellets of benzoic acid (which has an energy content of $26.45 \, kJ \, g^{-1}$) a straight-line calibration graph is compiled of the energy released in the bomb calorimeter against the corresponding movement of the recorder pen.

The bomb calorimeter described above is non-adiabatic, but some are adiabatic (which means that heat is not lost from the system). The latter can involve an inner and an outer water jacket. While combustion of the sample raises the temperature of the inner one, the outer is electrically heated so as to follow this temperature rise exactly. The rise in temperature of the inner jacket then precisely reflects the heat produced by the combustion, since no heat is lost to the surroundings.

It was pointed out in section 1.1 that a bomb calorimeter does not measure true enthalpy since the definition of change in enthalpy assumes constant pressure. The pressure in a bomb calorimeter during firing is far from constant but the error arising if heat of combustion is equated with enthalpy is less than 1 % (Wiegert, 1968).

The final point to be made is that calorific values obtained by bomb calorimetry are usually slightly lower than those estimated from chemical analysis. For example, Beukema and de Bruin (1979), using the tissues of the bivalve mollusc *Macoma*, found this discrepancy to be about 6 %. They point out that chemical methods can tend to produce over-estimates because commonly used conversion factors (e.g. 39.5 kJ g^{-1} for lipid, table 3.1) may be too high. Bomb calorimetry, on the other hand, can cause under-estimates through incomplete drying of the pellets or through incomplete combustion.

SUMMARY

1. Animals must derive all the energy they need from the food they eat (except for those which take in some nutrients through the skin). The energy content of a food material is indicated by its heat of combustion: for example 39.5 kJ g^{-1} for fat, 23.6 for protein and 17.2 for carbohydrate. Carnivores tap a rather more energy-rich food source than herbivores, because animal material generally has a higher energy content than plant material. Only part of the energy in the food can be put to use by the animal because, for example, some energy is lost in the faeces through incomplete digestion and absorption. Shortage of food may cause competition for it. Predators tend to feed more sporadically than herbivores and filter-feeders.

2. The factors that control the sensation of hunger and initiate feeding are subtle and only partially understood. In mammals the hypothalamus certainly plays a role. Regulatory mechanisms that have been proposed for mammals include responses to changes in the blood levels of glucose or fats, to changes in the rate of heat production, and to the extent of stomach distension. Sensory fatigue and gut distension, among other factors, are believed to cause insects to stop feeding.

3. The rate of food energy intake which exactly balances energy losses is known as the maintenance ration. A higher (super-maintenance) rate of intake is necessary if growth is to occur.

4. The metabolic rate of an animal generally rises for a time soon after feeding. This widespread phenomenon, commonly called specific dynamic action (SDA), can be demonstrated by monitoring changes in the rates of oxygen consumption, heat production, or nitrogenous excretion. The precise causes of SDA are still unclear and may vary with the species or other circumstances. In some cases degradation of "excess" amino acids seems responsible, in others protein synthesis may be the cause of the increased metabolic rate.

5. Energy intake can only be calculated when the energy values of samples of the food have been measured. The energy content of biological material is generally measured by bomb calorimetry. A sample is dried, weighed, placed in a thick-walled and air-tight steel chamber and completely burned by igniting it in a high pressure of oxygen. The heat released by the combustion is measured and compared with that produced by burning a substance of known energy content.

CHAPTER FOUR

ENERGY TRANSFORMATIONS WITHIN THE BODY

When a respiratory substrate such as glucose is completely broken down to carbon dioxide and water, molecules of adenosine triphosphate (ATP) are formed, which conserve some of the energy released in the process. The ATP can then give up energy to other molecules to do work within the body. Work may be done in the contraction of muscles, may be concerned in elaborating complex biochemicals (biosynthesis), or may enable ions and simple metabolites to be moved across cell membranes against the concentration gradient (active transport). This chapter is concerned with how energy is conserved in ATP synthesis, and how it is expended in muscle action, biosynthesis and active transport.

4.1 Adenosine triphosphate (ATP)

ATP is the molecule used to conserve energy derived from nutrient breakdown and to supply the energy needed in cellular processes. It thus links energy-liberating events with energy-utilizing ones. The structure of ATP is shown in figure 4.1. Its formation from adenosine diphosphate (ADP) and inorganic phosphate (P_i), and the reverse process in which energy is released from ATP for use, can be written

$$ADP + P_i \rightleftharpoons ATP + H_2O \qquad (4.1)$$

This is an over-simplification, however. The four hydroxyl groups (—OH) in the side chain of ATP (figure 4.1) are normally dissociated (i.e. in the form —O⁻). So under physiological conditions the molecule is negatively charged (but readily complexes with magnesium ions). A more accurate expression of the relation between ATP and ADP is

$$ADP^{3-} + HPO_4^{2-} + H^+ \rightleftharpoons ATP^{4-} + H_2O \qquad (4.2)$$

The bond which is broken when ATP is split to form ADP and

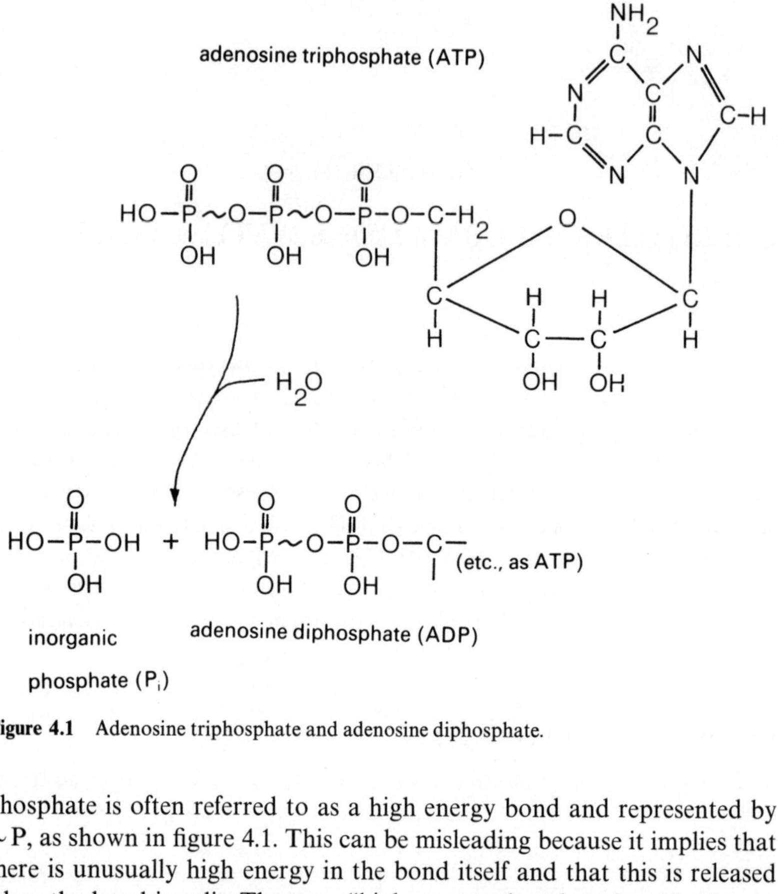

Figure 4.1 Adenosine triphosphate and adenosine diphosphate.

phosphate is often referred to as a high energy bond and represented by $\sim P$, as shown in figure 4.1. This can be misleading because it implies that there is unusually high energy in the bond itself and that this is released when the bond is split. The term "high energy phosphate bond" refers to the difference in energy content between ATP and its breakdown products, however, not to the bond energy of the phosphorus–oxygen link. The change in free energy when ATP splits to form ADP and phosphate is $-31 \, \text{kJ mole}^{-1}$ at 25°C and pH 7, the minus sign indicating that this energy is released.

Why, then, if the energy released by ATP breakdown is not exceptionally high, is ATP such a universal "currency" for dealing in energy exchanges in animals? The answer is that ATP lies about midway along the range of organic phosphates ranked according to their free energies of hydrolysis. Thus it is well placed to receive energy (in its formation) from those higher up and to release energy (in its breakdown) to those lower down. For example, the figure for diphosphoglycerate is $-49.3\,\text{kJ mole}^{-1}$, which is higher than that of ATP, and so when phosphoglycerate is formed from it ATP can also be produced:

$$\text{diphosphoglycerate}^{4-} + \text{ADP}^{3-} \rightarrow \text{phosphoglycerate}^{3-} + \text{ATP}^{4-} \qquad (4.3)$$

(This reaction occurs in the early stages of the respiration of glucose, as shown near the top of figure 4.2). Similarly, glucose-6-phosphate has a lower value than ATP, $-13.8\,\text{kJ mole}^{-1}$, and so ATP donates a phosphate in its production:

$$\text{glucose} + \text{ATP}^{4-} \rightarrow \text{glucose-6-phosphate}^{2-} + \text{ADP}^{3-} + \text{H}^{+} \qquad (4.4)$$

(This reaction is shown at the top of figure 4.2, for glucose-6-phosphate is an intermediate in the breakdown of glucose to fructose diphosphate.)

The value of $-31\,\text{kJ}$ for the change in free energy when ATP is broken down to ADP applies only under so-called standard conditions, that is to say at unimolar concentrations, pH 7 and 25°C. In living cells conditions may differ from these. The concentrations of ATP, ADP and phosphate are lower than 1 molar, the pH may not be 7, and the temperature will either vary (in poikilotherms) or be higher than 25°C (it is about 37°C in mammals). In addition, magnesium ions present in the cell form complexes with the reactants, shifting the equilibrium of the reaction. (In fact ATP cannot attach to an enzyme unless activated by magnesium ions.) The free energy of ATP breakdown to ADP and phosphate under physiological conditions is likely to be much higher than $-31\,\text{kJ}$ per mole, probably about $-50\,\text{kJ mole}^{-1}$.

4.2 Production of ATP

To see how ATP can be formed in the cells of an animal we will first consider the complete oxidation of glucose, to water and carbon dioxide, and the ATP production which results from it.

Glycolysis is the term given to the breakdown of glucose (or glycogen) to lactic acid, a process also known as the Embden–Meyerhof pathway. The lactic acid is formed from pyruvate when oxygen is lacking, and this process will be considered later. Here we are concerned with the situation when oxygen is available, in which case acetyl coenzyme A (acetyl CoA) is formed from the pyruvate. The sequence of events is outlined in the upper part of figure 4.2. The glucose is phosphorylated twice, by two ATP, forming fructose diphosphate. This is split into two and the products phosphorylated by two inorganic phosphates ($2P_i$), yielding two molecules

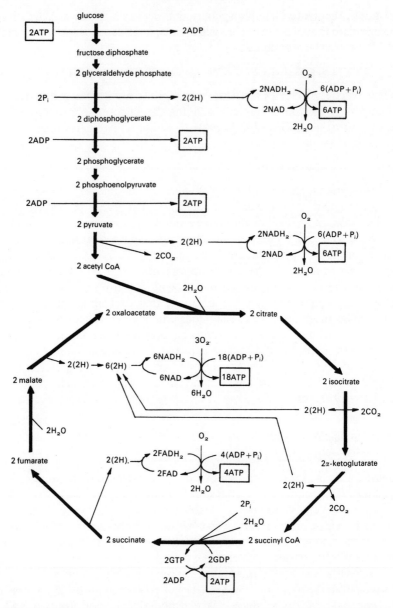

Figure 4.2 ATP production in the complete oxidation of one mole of glucose. 2 ATP are used but 40 ATP are formed, giving a net yield of 38 ATP. Not all the intermediates are shown.

of diphosphoglycerate (and two pairs of hydrogen atoms, whose fate will be considered later). In the formation of two pyruvates from these, four ATP are produced, giving a net gain of two. So the breakdown of glucose to pyruvate can be summarized as

$$C_6H_{12}O_6 + 2(ADP + P_i) \rightarrow 2\,C_3H_4O_3 + 2\,ATP + 2(2H) \tag{4.5}$$

The formation of pyruvate in this way occurs in the cytoplasm of the cell but the next stage, known as the *tricarboxylic acid cycle* (TCA cycle) or Krebs cycle, takes place within the mitochondria. These are double-membraned organelles, about $3\,\mu m$ long ($1\,\mu m$ is a millionth of a metre) and often sausage-shaped. The space between the outer and inner membranes of a mitochondrion is narrow ($0.01\,\mu m$ or less) but the matrix, enclosed by the inner membrane, is relatively spacious. The TCA cycle takes place in the matrix.

The TCA cycle is shown in the lower part of figure 4.2. Acetyl CoA, derived from the 3-carbon pyruvate, has only two carbon atoms. It combines with the 4-carbon oxaloacetate to give the 6-carbon citric acid. During the course of the TCA cycle oxaloacetate is eventually derived from citrate. Thus for each complete turn of the cycle from pyruvate, three carbon atoms are lost, as $3\,CO_2$. Since two pyruvates were produced from the original glucose, the six carbon atoms of the glucose appear as $6\,CO_2$. Of more significance, however, is the appearance of ten pairs of hydrogen atoms from the degradation of each pair of pyruvates. As shown in figure 4.2 these arise from the following steps: pyruvate to acetyl CoA, isocitrate to α-ketoglutarate, α-ketoglutarate to succinyl CoA, succinate to fumarate, and malate to oxaloacetate. As a result of the liberation of these hydrogen atoms much ATP is produced, as we shall shortly see. In addition, as succinate is produced from succinyl CoA, ATP is formed by way of guanosine diphosphate and triphosphate (GDP and GTP). Thus the breakdown of two pyruvates, in two turns of the TCA cycle, can be summarized as

$$2\,C_3H_4O_3 + 6\,H_2O + 2(ADP + P_i) \rightarrow 6\,CO_2 + 2\,ATP + 10(2H) \tag{4.6}$$

Adding together equations 4.5 and 4.6 gives

$$C_6H_{12}O_6 + 6\,H_2O + 4(ADP + P_i) \rightarrow 6\,CO_2 + 4\,ATP + 12(2H) \tag{4.7}$$

The fate of the twelve pairs of hydrogen atoms is shown in figure 4.2. Each of ten pairs reduces nicotinamide adenine dinucleotide (NAD) to $NADH_2$ (more accurately written $NADH + H^+$). Each of these ten pairs of hydrogens ends up as water, combining with an atom of oxygen, in a process which produces 3 ATP:

$$10(2H) + 5 O_2 + 30(ADP + P_i) \rightarrow 10 H_2O + 30 ATP \tag{4.8}$$

Each of the other two pairs of hydrogen atoms, produced in the step from two succinate to two fumarate, reduces flavin adenine dinucleotide (FAD) to $FADH_2$. Subsequent production of ATP is similar to that involving NAD, but in this case each pair of hydrogens results in the formation of only two ATP (not three, as when NAD is involved). Thus for those hydrogens involved with FAD

$$2(2H) + O_2 + 4(ADP + P_i) \rightarrow 2 H_2O + 4 ATP \tag{4.9}$$

Adding equations 4.8 and 4.9 gives

$$12(2H) + 6 O_2 + 34(ADP + P_i) \rightarrow 12 H_2O + 34 ATP \tag{4.10}$$

for oxidative phosphorylation involving the hydrogen released in the total breakdown of a mole of glucose.

We can now write an overall equation for the ATP yield from the complete oxidation of glucose, by adding equations 4.7 and 4.10:

$$C_6H_{12}O_6 + 6 O_2 + 38(ADP + P_i) \rightarrow 6 CO_2 + 6 H_2O + 38 ATP \tag{4.11}$$

This is the familiar equation for glucose respiration, with a net yield of 38 ATP. The overall free energy change for the oxidation of a mole of glucose is $-2870 \, kJ$. How much of this becomes conserved in ATP and how much is lost as heat? Assuming that each mole of ATP conserves $31 \, kJ$ under standard conditions (section 4.1), 38 will conserve $1178 \, kJ$. Thus the overall efficiency of the process, in terms of the amount of available energy conserved as ATP, is $(1178/2870) \times 100$, or 41%. Under conditions in the living cell, however, the free energy change may be $50 \, kJ \, mole^{-1}$ ATP (section 4.1) rather than 31, giving an efficiency of about 66%.

Of the twelve pairs of hydrogen atoms released during oxidation of a mole of glucose (figure 4.2), ten are liberated in the matrix of the mito-chondrion but one pair (that formed in the conversion of glyceraldehyde phosphate to diphosphoglycerate) is produced in the cytoplasm. How does the latter pair get into a mitochondrion, where oxidative phosphorylation occurs? There seems to be a "shuttle" in the mitochondrial membranes of liver and heart cells, involving malate and aspartate, which serves to take the two hydrogens from the $NADH_2$ in the cytoplasm and donate them to NAD in the mitochondrial matrix, forming $NADH_2$ there. By this method

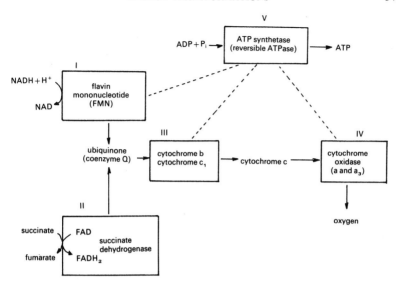

Figure 4.3 Schematic outline of the mitochondrial electron transport chain and oxidative phosphorylation. See text for explanation.

three ATP are generated in the matrix for each $NADH_2$ formed in the cytoplasm (top right of figure 4.2). In muscle cells the shuttle may be of a kind which takes hydrogens from cytoplasmic $NADH_2$ and causes production of $FADH_2$ from FAD in the mitochondrion. We have already seen that only two ATP are produced from each $FADH_2$ (as against three from each $NADH_2$), so in a system using such a shuttle only four ATP will result from the cytoplasmic two $NADH_2$. This will reduce the overall net yield in the oxidation of glucose from 38 ATP to 36 ATP.

The nature of the *oxidative phosphorylation* process remains to be considered. There is an electron transport chain, located in the inner mitochondrial membrane, which involves a series of cytochromes and other substances (figure 4.3). A cytochrome is a protein attached to a haem-like group containing a central iron atom. When ferric (Fe^{3+}) this iron can accept an electron and become ferrous (Fe^{2+}). At the beginning of the chain $NADH_2$ gives up its hydrogens:

$$NADH + H^+ \rightarrow NAD^+ + 2e^- + 2H^+ \tag{4.12}$$

The electrons ($2e^-$) are passed along the chain and some of the considerable fall in free energy that results is used to drive the formation of ATP from

ADP and phosphate. There is a total decrease in free energy along the length of the chain, from $NADH_2$ to oxygen, of about $-220\,kJ$. At the end of the chain the protons (H^+) and electrons combine with atomic oxygen to form water:

$$2e^- + 2H^+ + \tfrac{1}{2}O_2 \rightarrow H_2O \qquad (4.13)$$

The components of the electron transport chain are in complexes within the membrane. The more important functional components of these are shown in figure 4.3. Complexes I to III also contain "iron-sulphur proteins" (usually written Fe-S) and IV contains copper. Complex II contains the only TCA cycle enzyme which is not free in the matrix, succinate dehydrogenase. It was once thought that ATP was directly generated in complexes I, III and IV, which helped to explain how three ATP are formed for each $NADH_2$ but only two ATP for each $FADH_2$ (see figure 4.3). The situation is much more complicated than this, however, and several ideas of the true mechanism are current (see Jones, 1981, for a concise review). What probably happens, in essence, is that protons are ejected from the outer surface of the inner mitochondrial membrane into the intermembrane space. These protons then pass back across the inner membrane and in so doing drive the phosphorylation of ADP to ATP in a fifth complex, comprising units F_0 and F_1 and shown as V in figure 4.3. The F_1 units seem to be located in minute knobs on the inner surface of the inner membrane.

So far we have considered only the ATP yield from the oxidation of glucose, but amino acids and fatty acids can also be respired. Figure 4.4 shows some ways in which these can be involved in the TCA cycle. When carbohydrate is plentiful it can be either respired or stored (as glycogen), or it can be converted into fatty acids by way of pyruvate and acetyl CoA. When carbohydrate is in short supply, fatty acids can provide the acetyl CoA input into the TCA cycle. Fatty acids with an even number of carbon atoms can be completely degraded into acetyl CoA, one acetyl CoA being produced for each pair of carbon atoms. Most of this process, known as β-oxidation, takes place in the mitochondria. In the case of oxidation of the 6-carbon fatty acid, caproic acid, 3 acetyl CoA will be produced. As two turns of the TCA cycle from acetyl CoA produce 24 ATP (figure 4.2), three turns will yield 36. In addition, there is a net gain of 8 ATP during the β-oxidation of caproic acid to 3 acetyl CoA, giving a total net gain of 44 ATP from the degradation of the 6-carbon caproic acid—more than the net gain of 38 ATP arising from the breakdown of glucose (equation 4.11), also a 6-carbon compound.

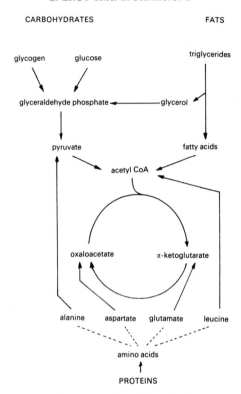

Figure 4.4 Some pathways whereby protein and fat derivatives can enter the TCA cycle. The fate of carbohydrates (shown in more detail in figure 4.2) is also indicated.

We have seen that, for ATP production by the electron transport chain and oxidative phosphorylation, oxygen is required to accept hydrogen ions and form water. If oxygen is not available, anaerobic metabolism which can produce ATP will be necessary. In a hard-working muscle, for example, the oxygen supply may be insufficient to allow the TCA cycle and oxidative phosphorylation to occur, in which case the breakdown of glucose to lactic acid (glycolysis) by way of pyruvate must suffice. (The lactic acid may accumulate and be oxidized when oxygen is again available.) The overall equation for glycolysis in which glucose breakdown results in the production of two molecules of lactic acid is as follows:

$$C_6H_{12}O_6 + 2(ADP + P_i) \rightarrow 2\,CH_3 \cdot CHOH \cdot COOH + 2\,ATP \qquad (4.14)$$

There is a net gain of only 2 ATP, but then there is an overall change in free energy of only $-197\,\text{kJ}$ in the production of lactate from glucose. Assuming one ATP conserves $31\,\text{kJ}$ (section 4.1) two will conserve $62\,\text{kJ}$, and the efficiency of the process is $(62/197) \times 100$, or $31.5\,\%$. The figure calculated above for the complete breakdown of glucose, with a net gain of 38 ATP, was $41\,\%$. So although glycolytic production of lactate yields much less ATP per mole of glucose, it is not very much less efficient in terms of the proportion of the available energy which becomes trapped as ATP.

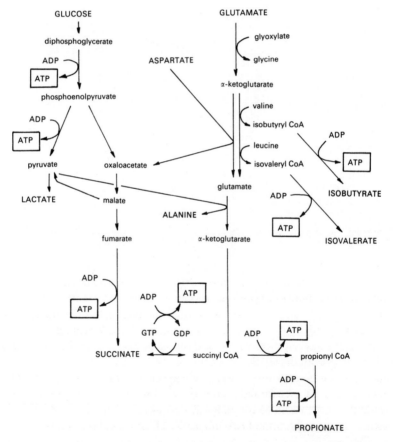

Figure 4.5 An outline of the pathways which can produce ATP anaerobically. Some intermediates are omitted. (Based on Hochachka *et al.*, 1973, and Hochachka and Somero, 1976.)

Many invertebrate animals live in habitats which are deficient in oxygen, at least at times, and such forms show a variety of metabolic pathways which generate ATP in the absence of oxygen. Among the animals which have been studied in this respect are the intertidal mussel *Mytilus* and the gut parasite *Ascaris*. It is not yet clear which animals resort to which pathways, but the options open to invertebrates needing to respire anaerobically are summarized in figure 4.5. Glutamate and aspartate are used as substrates and the end-products include alanine, succinate and propionate, with smaller amounts of isobutyrate and isovalerate. The production of lactate from glucose is shown in the upper left part of figure 4.5, but although this anaerobic source of ATP predominates in vertebrates it is often of minor importance among invertebrates, and in some cases lactate is not produced at all. Although there are numerous points of origin of ATP among the pathways shown in figure 4.5, power output is relatively low (in terms of ATP yield per unit time) compared with the rates which generally result from glycolysis in vertebrate muscle.

4.3   Muscle contraction

Most animals move about in order to find food or avoid capture, or to find a mate. Movements are possible because muscles (or other contractile elements, such as cilia and flagella) can transform chemical energy into mechanical work, potential energy into kinetic energy. The energy is obtained from the breakdown of ATP and the work is done as the force exerted by the contracting muscle is transmitted to other objects. No work results as the muscle relaxes, so muscles can only pull, not push. Many are connected to skeletal elements, however, forming a system of levers which allows a wide variety of directional forces to be applied.

The proteins actin and myosin are involved in the contraction mechanism of all muscles but their arrangement varies according to the muscle type. In striped muscle (voluntary, or skeletal muscle) each fibre, which is a syncytium of several cells, contains many myofibrils. A fibril consists of repeating units (sarcomeres) of bundles of actin filaments interdigitating with bundles of myosin filaments. When the muscle contracts, these filaments do not shorten appreciably but slide further between one another (figure 4.6). Myosin filaments are of fairly constant length among the vertebrates, but the actin filaments are more variable. A myosin filament is made up of many molecules, each shaped like a golf club with the head pointing outwards towards an actin filament. The latter is a

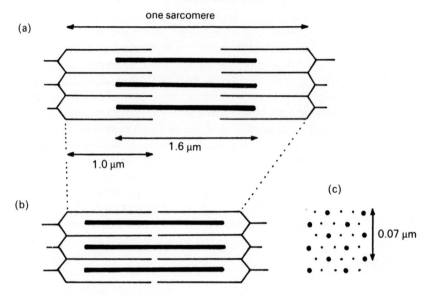

Figure 4.6 The arrangement of actin (thin filaments) and myosin (thick filaments) in a relaxed (*a*) and contracted (*b*) sarcomere from striped muscle. (*c*) Transverse view of the region of actin and myosin overlap.

twisted double strand of spherical sub-units, and lying in the grooves between the strands are long molecules of tropomyosin. At regular intervals (about $0.04\,\mu m$) along the actin filament are small molecules of yet another protein, troponin. The arrival of a motor nerve impulse at the muscle surface causes a rise in the Ca^{2+} concentration around the myofibrils. Calcium binds to the troponin, and this probably has the effect of shifting slightly the position of the tropomyosin relative to the actin. The head of a myosin molecule can then bind temporarily to the actin, as shown in figure 4.7*b*. The myosin head releases ADP and P_i into the sarcoplasm and also bends, as indicated in figure 4.7*c*, moving the thin filament by about $0.01\,\mu m$ relative to the thick one. The myosin head takes up ATP and the bridge between myosin and actin is broken (figure 4.7*d*). The ATP is hydrolysed into ADP and P_i and the energy released is used to force the head into the "strained" position shown in figure 4.7*a*. The cycle can repeat if Ca^{2+} remains bound to the troponin and if the ATP supply is adequate.

The demand for ATP in a working muscle is high and most animals have a phosphagen, such as creatine phosphate (figure 4.8), which can

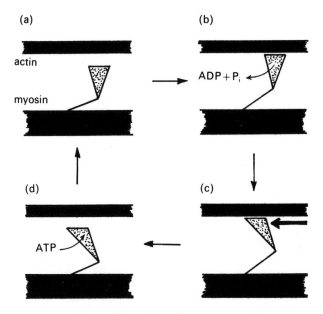

Figure 4.7 An outline of the making and breaking of a cross-bridge between actin and myosin. (Based on Wilkie, 1976.)

serve as a store of high-energy phosphate and donate it to ADP when required:

$$\text{creatine phosphate} + \text{ADP} \rightarrow \text{creatine} + \text{ATP} \qquad (4.15)$$

Thus during muscle contraction there is a sharp fall in creatine phosphate, but when the muscle is relaxed stocks are renewed, from ATP formed in the mitochondria. Creatine phosphate is found in all vertebrates and also in a few invertebrates, but arginine phosphate (figure 4.8) is characteristic

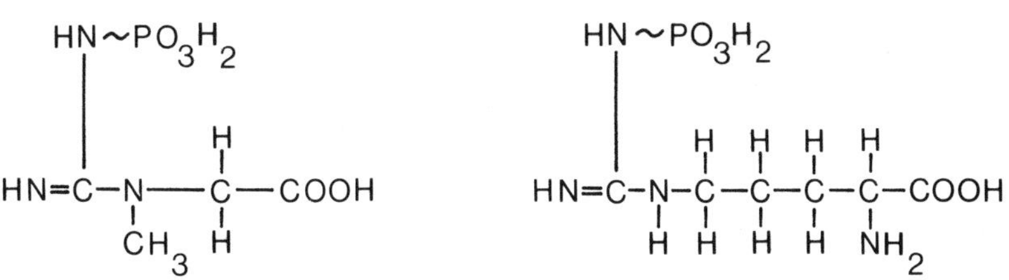

Figure 4.8 Creatine phosphate (left) and arginine phosphate.

of the latter. In addition to these two widespread phosphagens, several others are known to occur among certain invertebrate groups (for example the annelids).

Striped muscle fibres generally contract faster than those of smooth muscle (involuntary, or visceral muscle). Some insect flight muscles can contract in less than a millisecond. Striped muscles shorten by about 30%, but smooth muscles can shorten by up to 80% of their resting length. If the tension exerted by a muscle is constant as it shortens, the contraction is said to be *isotonic*, whereas if no appreciable shortening occurs the contraction is *isometric*. In an isometric contraction, energy utilization and heat production are high at first but rapidly fall. More energy is used and more heat is produced in an isotonic contraction, in which the heat production is greater, in fact, than can be accounted for by ATP or creatine phosphate breakdown. Perhaps some unknown reactions are taking place (Goldspink, 1977). The maximum force that a muscle can generate is usually 10 to 40 newtons per cm^2 cross-sectional area (a newton is the force that gives 1 kg an acceleration of 1 metre sec^{-2}), but some molluscan muscles can produce 140 newtons per cm^2. A mole of creatine phosphate produces about 46 kJ (Wilkie, 1976), and in a fast muscle of a hamster, for example, about 2 micromoles of high energy phosphate per gram are used in a 30-second isometric contraction (Goldspink, 1977).

Much of the energy made available to a muscle is lost as heat, the proportion varying with the type of muscle and the conditions of contraction. The efficiency of the contractile process itself is probably about 80 to 90% in frog muscle, but if ion-pumping and the whole process of contraction and recovery are taken into account the overall efficiency is only about 20 to 25% (less than that of diesel engines, for example, which have an efficiency of about 40%). The work and heat produced by muscles have recently been reviewed by Curtin and Woledge (1978), where more information may be found. To keep the muscles functioning effectively, energy must also be expended on support systems. Ventilation of gills or lungs, and maintenance of heartbeat, are energy-demanding processes necessary if the muscle is to receive an adequate supply of oxygen and respiratory substrate.

4.4 Biosynthesis

Energy is expended in cells during the synthesis of complex biochemicals. Monosaccharides are assembled into polysaccharides, amino acids into proteins, fatty acids into triglycerides and phospholipids, and nucleotides into the nucleic acids DNA and RNA.

Animals elaborate the polysaccharide glycogen mainly in liver and muscle tissue. Glucose residues are joined together by glycosidic links into branched chains. The formation of each glycosidic bond requires two molecules of ATP (so the formation of a glycogen molecule of 10 000 glucose residues requires 20 000 ATP). One is used to phosphorylate the glucose to glucose-6-phosphate and the other donates a high energy phosphate to uridine diphosphate, forming uridine triphosphate (UTP). The UTP is involved in linking the glucose residue to the growing glycogen chain. The free energy of hydrolysis of a glycosidic bond is about 17 kJ under standard conditions, and so the energy released by the two ATP (62 kJ, section 4.1) is ample for the purpose of forging the bond.

A protein is a chain of amino acid residues linked together by peptide bonds in a complicated process involving nucleic acids. An amino acid is first attached to its appropriate transfer RNA, and this requires the energy from two high energy phosphate bonds. (Both high energy bonds of an ATP molecule are broken, producing adenosine monophosphate.) Another ATP donates a high energy phosphate to guanosine diphosphate, forming guanosine triphosphate (GTP). Energy from subsequent GTP breakdown is used to move the growing peptide chain (and the messenger RNA which dictates its amino acid sequence) relative to the ribosome. (For details of the complexities of protein synthesis see, for example, White et al., 1978.) Thus the formation of one peptide bond, which incorporates about 23 kJ, requires three high energy phosphate bonds, so building a polypeptide of 150 amino acid residues will use 450 such bonds. The efficiency of energy utilization in peptide bond formation is $23/(3 \times 31)$, or 25 %.

In the case of fats, Lehninger (1971) has considered the energy locked in new bonds during the synthesis of a common phospholipid, phosphatidyl ethanolamine. In a triglyceride three fatty acids are linked to glycerol ($CH_2OH \cdot CHOH \cdot CH_2OH$), but here a phosphate and ethanolamine (a serine derivative) replace the third fatty acid. Seven ATP are used in the synthesis and the free energy input is about 71 kJ (14.6 kJ for joining each of the two fatty acids to the glycerol, 14.6 kJ for the phosphate-ethanolamine link, and 27.2 kJ for the phosphate-glycerol bond). The efficiency of the synthesis is therefore $71/(7 \times 31)$, or 33 %. Cytidine triphosphate (CTP) is involved in the process, whereas UTP is involved in glycogen synthesis and GTP in that of protein.

Other macromolecules requiring energy for their synthesis are the nucleic acids DNA and RNA. Two high energy phosphate bonds are needed to join two nucleotides in DNA. In addition, some work must be expended in organizing the superstructure of cells (membranous inclusions and the like) but such aspects have received little attention from the point of view of energetics.

What is the work rate required in the cell to provide the types of macromolecular synthesis outlined above? Information is sparse, but Lehninger (1971) calculates that in the bacterium *Escherichia coli* there are about 15 million lipid molecules, nearly 2 million protein molecules and about 40 000 polysaccharide molecules. Synthesis of a lipid from free fatty acids requires only about 7 molecules of ATP, however, whereas making a protein requires about 1500. In *E. coli* over 2 million ATP molecules are involved in protein synthesis every second, and this process is responsible for 88 % of the total energy used in biosynthesis (the synthesis of lipids requiring about 4%, polysaccharides about 3% and nucleic acids about 5%). At the opposite extreme of evolutionary complexity, man also contains far more lipid than protein or polysaccharide. White *et al.* (1978) state that an average (70 kg) man contains about 15 kg of triglycerides, 6 kg of protein (in muscle) and about 0.2 kg of glycogen (mainly in the liver and muscles). These represent energy reserves of about 590 000 kJ, 100 000 kJ and 3500 kJ respectively.

There is one further point to make about the energy cost of biosynthesis. In energy budgets where the energy lost as heat (R) is calculated from measurements of oxygen consumption, the energy of the bonds made in constructing the macromolecules discussed above is double-counted, for it is included in the estimation of R as well as in the measurement of the energy of growth materials (P). The resulting error is small, but why it arises is discussed in section 5.1.

4.5   Active transport

Ions and simple metabolites diffuse across cell membranes down concentration gradients. Yet cells tend to maintain within themselves constant concentrations of such materials (an aspect of homeostasis) even in the face of widely different concentrations outside. This steady state is achieved by active transport processes. In active transport a carrier molecule moves the ion or metabolite for which it is specific across the cell membrane against the concentration gradient. This requires the expenditure of energy, which is provided by ATP. Intestinal cells take up amino acids and glucose from the contents of the gut lumen by active transport. Most cells contain much lower concentrations of Na^+ and much higher concentrations of K^+ than those outside. These differences are established and maintained by an active transport process in the cell membrane which ejects Na^+ from the cell and carries K^+ into it. The mechanisms of active transport, not yet fully understood, have been concisely reviewed by Harrison and Lunt (1980) in another book in this series. It is perhaps surprising that animals have not evolved an active transport mechanism for oxygen, which is always taken up by simple diffusion.

Considerable concentration differences are often established by active transport. In a giant fibre of a squid (a system much studied by neurophysiologists) the concentration of potassium is twenty times that outside the cell. Perhaps the most remarkable case is the active transport of hydrogen ions into the stomach lumen to establish a low pH there, for the cells involved are moving out hydrogen ions against a concentration gradient of about ten million to one.

The energy required by the active transport of uncharged solutes such as glucose can be calculated fairly easily. Lehninger (1971) gives as an example the transport of a mole of glucose against a concentration gradient of 100 to 1 (e.g. from a concentration of 0.001 molar to one of 0.1 molar) and finds the minimum energy requirement for this to be 11.3 kJ. (Considering that the heat of combustion of a mole of glucose is 2833 kJ, this is a small price to pay.) Uptake of glucose by intestinal cells seems to be associated with sodium transport. The carrier in the cell membrane which brings glucose into the cell transports Na^+ simultaneously. An ATP-requiring sodium pump then ejects the Na^+ from the cell, to maintain the sodium concentration gradient. Active uptake mechanisms for various amino acids (for example alanine, glycine and tyrosine) can also be linked with sodium transport.

Sodium and potassium transport are combined in the so-called Na^+-K^+-ATPase system, which probably comprises two or more protein molecules located in the cell membrane. A high proportion of the basic metabolic rate of an animal is thought to be due to the cost of operating this system. In the case of erythrocyte cell membranes the pump expels three sodium ions from the cell and carries two potassium ions inwards for each molecule of ATP used. During the passage of an action potential along the axon of a neurone, Na^+ leaks into the cell and K^+ leaks out, these movements occupying about a millisecond. Eventually active transport moves the Na^+ back out and the K^+ back in, but this need not be invoked for some time, as the numbers of ions that move in and out of the cell during an action potential are a tiny proportion of those present. It has been calculated that a squid giant axon can conduct half a million action potentials before it needs to call on active transport to restore the high external N^+ and internal K^+ concentrations.

Active transport of Na^+ is also important in the kidney. In the loop of Henlé of a mammalian nephron, Na^+ is actively moved back across the wall of the ascending limb from the formative urine. The high Na^+ concentration which results draws water by osmosis from the nearby collecting tubules. This avoids undue loss of water in the urine.

Controlling the ionic concentration and water content of the tissues is an important aspect of homeostasis for all animals, but the problem is particularly acute for estuarine and freshwater forms, where for some or all of the time the animal is in an environment with an ionic concentration much lower than that of its body fluids, so that water enters the body by osmosis and ions leave it by diffusion. Active uptake of ions maintains the high internal concentrations. In a freshwater fish such uptake occurs mainly at the gills, but ions are also recovered in the kidney to reduce urinal loss. (The urine is copious and dilute, removing water which has entered the body by osmosis.) The energy used by estuarine and freshwater animals in taking up ions against the gradient is only a small proportion of total energy expenditure, however. Potts (1954) calculated that at 15°C a 60 g crayfish (*Potamobius*) expends about 0.15 joules an hour in maintaining the ionic concentration of its blood seventy-fold higher than that in the water outside. This is less than 0.4 % of the total energy expenditure. Similar calculations for the crab *Eriocheir* and the freshwater bivalve *Anodonta* yielded values of about 0.5 % and 1.2 % respectively. The crab *Carcinus*, which can live in estuaries as well as in fully marine intertidal habitats, has to allocate about 0.6 % of its total energy expenditure to "osmotic work", and the estuarine polychaete *Nereis diversicolor* about 0.4 %. These estimates of the energy expended in maintaining concentration differences by active transport are derived from calculations involving such factors as the external and internal concentrations, the total area of permeable body surface and the permeability of this surface. They are estimates of the minimum energy expenditure involved, for they take no account of any inefficiency in the uptake mechanisms, and so the actual work rates may be much higher.

SUMMARY

1. Adenosine triphosphate (ATP) is the molecule used to conserve energy derived from nutrient breakdown and to supply the energy needed in cellular processes. It thus links

energy liberation with energy utilization. Energy is made available when ATP splits into adenosine diphosphate and inorganic phosphate. The free energy of this breakdown, under physiological conditions, is probably about $-50 \, \text{kJ mole}^{-1}$.

2. Much ATP is produced during the complex sequence of biochemical events in which glucose is completely broken down. Parts of this process, the TCA cycle and oxidative phosphorylation, occur in the mitochondria. Cellular respiration of fatty acids or amino acids also produces ATP. If oxygen is lacking, ATP can be formed during the breakdown of glucose to lactic acid, or in a variety of processes known to occur in certain invertebrates accustomed to oxygen lack.

3. ATP supplies the energy involved in muscle contraction. In striped muscle, arrays of actin filaments interdigitate with myosin filaments. Other proteins (tropomyosin and troponin) are also involved, but in essence what happens in contraction is that the myosin and actin filaments slide further between one another as a result of the formation of temporary cross-bridges. Substances such as creatine phosphate and arginine phosphate can provide an energy store to maintain the supply of ATP. The maximum force that a muscle can generate is usually 10 to 40 newtons per cm^2 of cross-sectional area. The overall process of muscular contraction and recovery has an efficiency of only about 20 to 25 %, much energy being lost as heat.

4. A second major use of ATP is in the synthesis of macromolecules. Energy is expended in linking a chain of glucose residues together to form the polysaccharide glycogen, for example, and in linking amino acids together to form a protein. In the case of protein, the building of a peptide of 150 amino acid residues requires 450 high energy phosphate bonds. Energy is used in peptide bond formation with an efficiency of about 25 %.

5. Ions and simple metabolites are moved across cell membranes against concentration gradients by energy-requiring processes. This active transport maintains steady concentrations within the cells. The minimum energy requirement to transport a mole of glucose against a concentration gradient of 100 to 1 is about 11 kJ. Three sodium ions are pumped out of an erythrocyte and two potassium ions carried inwards for each molecule of ATP used. Active transport of Na^+ and K^+ is also an important part of neuronal activity and of kidney function. Estuarine and freshwater animals take up ions actively to combat loss by diffusion, but this "osmotic work" represents only a small percentage of their total energy expenditure.

CHAPTER FIVE

RESPIRATORY RATE AND THE RATE OF HEAT LOSS

Energy lost as heat is represented by R in an energy budget where $C = P + R + U + F$ (equation 2.2). It can be measured either directly or indirectly, as explained below. The respiratory rate, or metabolic rate, must be sufficient to supply enough ATP for muscular activity and various other energy-requiring processes. The heat which results from these is lost from the body. Homeotherms are a special case in this respect, as heat production and heat loss are balanced and the body temperature remains constant.

5.1 Direct and indirect calorimetry

Direct calorimetric methods measure the heat lost by an animal, whereas indirect methods generally monitor oxygen consumption and multiply this by a suitable coefficient to arrive at an estimate of heat production.

Direct calorimetry began two hundred years ago, when Lavoisier and Laplace enclosed a guinea pig in a chamber surrounded by ice. As the animal produced and lost heat, ice melted and the resulting water drained away. Knowing the amount of water produced and the latent heat of melting ice (335 joules g^{-1}) they were able to calculate the rate of heat loss, which was about 12.4 kJ an hour. A separate outer jacket, containing a mixture of ice and water, ensured that all the heat lost by the animal melted ice in the inner jacket. (No heat could pass to the outside as the inner and outer walls of the outer jacket were both at 0°C.)

Since the days of Lavoisier many direct calorimeters have been devised, of ever-increasing accuracy and complexity (see Kleiber, 1950, for a full review of the earlier ones). They are used to measure the heat produced by chemical reactions as well as that by whole animals. With the advent of thermistors and other modern technology it has become possible to measure heat loss from aquatic animals as well as from mammals. Thermistors are ceramic semi-conductors made by sintering mixtures of various metal oxides such as those of copper, iron, manganese and nickel. As the temperature of a thermistor rises its electrical resistance falls and so, after calibration, the temperature of a thermistor can be deduced by measuring its resistance. Thermocouples are also used extensively in direct calorimeters (and in bomb calorimeters, section 3.5). When two wires of different materials, for example copper and constantan (an alloy of copper and nickel), are joined together at both their ends, and one junction is hotter than the other, a current flows. The current can be measured and is directly dependent on the temperature difference between the junctions. Thus thermocouples can be used as thermometers, and when large numbers of them are built into an array in the wall of a direct calorimeter they provide a method of measuring the rate of heat transfer through the wall.

Although of bewildering variety, direct calorimeters are of two basic types. In adiabatic calorimeters heat is not passed to the surroundings, whereas in conduction calorimeters it is. Lavoisier and Laplace's calorimeter was a simple example of the adiabatic type, but conduction calorimeters have found wide application in biology. In a conduction calorimeter heat flow across the wall can be measured (by thermocouples, for example, as mentioned above) or the heat can be taken up by piped circulating water.

The design of direct calorimeters for large air-breathing animals such as mammals is continually being improved. An air supply to the subject must be maintained and account must be taken of evaporative heat loss, from the lungs and skin, as well as "sensible heat loss" through convection and radiation. (The latent heat of vaporization of water varies with temperature, which provides a further complication. It is about 2.41 kJ g^{-1} at 37°C, the body temperature of mammals.) Studies of human nutrition and energy balance involve the use of direct calorimetry (Jéquier, 1977), and modern agricultural methods incorporate information from direct calorimetry with cattle (Blaxter, 1971) and poultry (Farrell, 1974). In addition, it has recently become possible to build direct calorimeters for use with a variety of aquatic invertebrates, chiefly annelids, gastropod and bivalve molluscs, and crustaceans (see Gnaiger, 1980; Hammen, 1980; Lowe, 1978; Pamatmat, 1979). Such studies have tended to involve indirect calorimetry as well, and will be returned to later. The direct calorimeter built by Lowe (1978), primarily for use with a polychaete, is shown in figure 5.1. Water flow past the worm is constant and continuous,

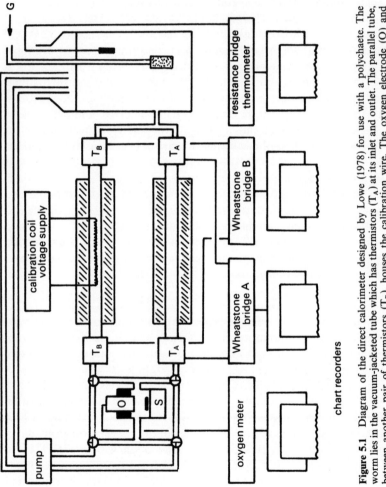

Figure 5.1 Diagram of the direct calorimeter designed by Lowe (1978) for use with a polychaete. The worm lies in the vacuum-jacketed tube which has thermistors (T_A) at its inlet and outlet. The parallel tube, between another pair of thermistors (T_B), houses the calibration wire. The oxygen electrode (O) and magnetic stirrer (S) allow simultaneous indirect calorimetry. Air or other gas mixtures can be bubbled (by way of G) through the stock water, whose temperature is monitored by the resistance bridge thermometer.

and thermistors at the entrance and the exit of the chamber are incorporated in a Wheatstone bridge in such a way that the temperature difference between them, due to heat loss by the animal, can be displayed on a strip-chart recorder. Calibration is achieved by passing a current through a wire coil of known heat production, in a twin chamber. Oxygen consumption is also continuously measured. Advantages of this system are the continuous flow (the animal is in a constant ambient oxygen concentration) and the fact that heat loss and oxygen uptake are measured simultaneously and continuously.

Measurement of oxygen consumption forms the basis of most methods of indirect calorimetry. A polarographic oxygen electrode is often used nowadays, which consists of a small pair of metal electrodes separated by an electrolyte and enclosed by a membrane through which oxygen diffuses. When a suitable voltage is put across the electrodes oxygen becomes dissociated and a current flows. If the supply of oxygen molecules to the

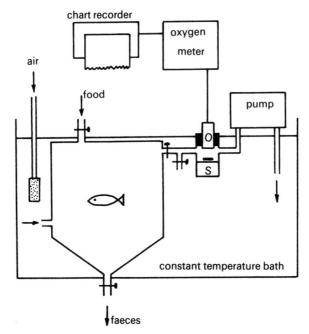

Figure 5.2 A continuous-flow respirometer of the type used by Solomon and Brafield (1972) to compile an energy budget for a fish (see Figure 2.3). The magnetic stirrer (S) allows the oxygen electrode (O) to monitor either the oxygen concentration of the water which has been through the respiration chamber or, by manipulating screw clips, that of the stock water.

system is kept up (for example by stirring the medium being tested) the current is proportional to the concentration or partial pressure of oxygen. It can be amplified and displayed on a strip-chart recorder. With vertebrates and aquatic invertebrates, such oxygen electrodes can provide a continuous record of the oxygen content of the air or water which has been pumped through the chamber containing the animal, and so its rate of oxygen consumption can be calculated. Continuous-flow respirometry is preferable to following the decline in oxygen concentration in a closed system, as the animal is in a constant environment from the point of view of oxygen availability, and oxygen uptake can therefore be monitored over long periods. In the respirometer shown in figure 5.2, for example, oxygen consumption can be measured continuously and accurately for a month or so (during which time significant growth will have occurred) and used to calculate the energy lost as heat (R). The energy of the food consumed (C) and of the faeces produced (F) can also be determined. If growth and the rate of nitrogenous excretion are estimated, the energy of growth (P) and of nitrogenous waste (U) can be calculated, and so a complete energy budget can be compiled.

To convert measurements of oxygen consumption into estimates of the energy lost as heat, an oxycalorific coefficient (Q_{ox}) must be applied. This varies according to the respiratory substrate. The respiration of one mole of glucose to carbon dioxide and water requires 6 moles of oxygen (192 g) and releases 2833 kJ as heat (table 3.1), so the appropriate Q_{ox} for carbohydrate is 2833/192 or 14.76 joules mg^{-1} oxygen consumed. The Q_{ox} for fats varies according to their composition but the average value is generally agreed to be 13.72 joules mg^{-1} oxygen.

The situation for protein is less clear, as the Q_{ox} varies with the nitrogenous excretory product. Most aquatic invertebrates and freshwater fish excrete predominantly ammonia, but in mammals and some other

Table 5.1 Oxycalorific coefficients as joules lost as heat mg^{-1} oxygen consumed (Q_{ox}) and joules lost in nitrogenous excretory products mg^{-1} oxygen consumed in respiring protein (Q_{ex}).

	Q_{ox}	Q_{ex}
carbohydrate (glucose)	14.76	—
fats (mean value)	13.72	—
protein to ammonia	13.36	2.70
protein to urea	13.60	2.46
protein to uric acid	13.60	4.12

groups urea is the chief excretory product. Insects, birds and terrestrial reptiles produce uric acid (chapter 7). Consider the respiration of 100 g of protein consisting of 53 % carbon, 7 % hydrogen, 23 % oxygen and 16 % nitrogen. (The remaining 1 % is sulphur and will be ignored as the amount is so small and the form in which it is excreted is uncertain.) After dividing these percentages by the appropriate atomic weights (12, 1, 16 and 14 respectively) the following balanced equations can be compiled:

$$+4.6\,O_2 \;\rightarrow\; 1.14\,NH_3 \qquad +4.42\,CO_2 + 1.79\,H_2O \quad (5.1)$$

4.42 C (4.6 × 32 = 147.2 g) (ammonia at 347.9 kJ mole^{-1} = 397 kJ)

7.00 H

$$+4.6\,O_2 \;\rightarrow\; 0.57\,CO(NH_2)_2 \;+3.85\,CO_2 + 2.36\,H_2O \quad (5.2)$$

1.44 O (4.6 × 32 = 147.2 g) (urea at 634.3 kJ mole^{-1} = 362 kJ)

1.14 N

(100 g protein, energy value 2364 kJ)

$$+4.168\,O_2 \;\rightarrow\; 0.285\,C_5H_4O_3N_4 + 2.995\,CO_2 + 2.93\,H_2O$$

(4.168 × 32 = 133.4 g) (uric acid at 1926 kJ mole^{-1} = 549 kJ) $\qquad$ (5.3)

From equation 5.1 it can be seen that the oxycalorific coefficient when ammonia is being produced is $(2364 - 397)/147.2$, or 13.36 joules released as heat mg^{-1} oxygen used in respiring protein. By similar calculations the Q_{ox} for urea production is found to be 13.60 joules mg^{-1} oxygen (from equation 5.2), and the same value is appropriate for uric acid production (equation 5.3). (The respiratory quotients, or ratios of carbon dioxide produced to oxygen consumed, are 0.96, 0.84 and 0.72 for ammonia, urea and uric acid production respectively.)

It is interesting to note here that equations 5.1 to 5.3 also allow the energy lost in the excretory products (U in the energy budget equation) to be calculated in terms of oxygen consumption (Brafield and Solomon, 1972). Thus $397/147.2$ or 2.7 joules are lost in ammonia per mg oxygen consumed in respiring protein. Such calculations are useful when compiling energy budgets in situations where the quantities of nitrogenous wastes produced are difficult to measure accurately. The oxycalorific coefficients calculated above, and the energy lost in excretory products in terms of oxygen consumed in respiring protein, are summarized in table 5.1. The values for protein in this table are almost identical to those calculated by Elliott and Davison (1975) who, rather than ignoring the small amount of sulphur in protein, assumed it to be released as ammonium sulphate. They also used the energy value for aqueous urea, rather than that for solid urea shown in equation 5.2.

Unfortunately, one can rarely assume that an animal is respiring a single substrate, and so a composite Q_{ox} has to be calculated. Thus if carbohydrate, fat and protein are being respired in the ratio $7:2:1$ the appropriate Q_{ox} (if the animal is producing ammonia) will be

$$(0.7 \times 14.76) + (0.2 \times 13.72) + (0.1 \times 13.36)$$

or 14.41 joules mg^{-1} oxygen consumed. One may be forced to assume, however, that substrates are being respired in the proportions in which they are being ingested, or assimilated. This is a dangerous assumption, unlikely to be valid, and so inaccuracy can arise in calculating the oxycalorific coefficient.

In the case of mammals, which produce urea, it is fairly easy to measure oxygen consumption, carbon dioxide production and urea nitrogen production. The equation derived by Weir (1949) can then be applied:

$$R = (3.941 \times O_2) + (1.106 \times CO_2) - (2.17 \times N) \tag{5.4}$$

where R is the total heat output in kcal (kJ if multiplied by 4.184), oxygen consumption and carbon dioxide production are in litres, and N is g of urine nitrogen. Ruminants, such as cows, produce methane as a result of stomach fermentation, which is lost through the mouth. Here the equation of Brouwer (1965) can be applied:

$$R = (3.866 \times O_2) + (1.200 \times CO_2) - (0.518 \times CH_4) - (1.431 \times N) \tag{5.5}$$

(Symbols are as in equation 5.4 and methane production is in litres.) For studies on birds (usually poultry), which produce uric acid, Farrell (1974) has derived the equation

$$R = (16.20 \times O_2) + (5.00 \times CO_2) - (1.20 \times N) \tag{5.6}$$

where R is the heat production in kJ and N is g of nitrogen in the excreted uric acid. Now that it is feasible to measure accurately carbon dioxide production (with electrodes) and ammonia production (chemically) by aquatic animals, it should soon be possible to apply an equation similar to those above to ammonioteles. One of us has derived the following equation, partly based on the procedure used by Weir (1949):

$$R = (11.16 \times O_2) + (2.62 \times CO_2) - (9.41 \times NH_3) \tag{5.7}$$

where R is heat loss in joules, and oxygen consumption, carbon dioxide production and ammonia production are in mg.

Another source of error in using indirect calorimetry for calculating the energy lost as heat (R) for the purpose of compiling an energy budget is the fact that a small amount of energy is double-counted. An oxycalorific coefficient assumes that all the energy released in the respiratory process is lost as heat, but in a growing animal this is not the case—some energy from the ATP which has been generated is not lost as heat but is put into, and retained in, the synthesized macromolecules (some has been conserved as potential energy). The energy in the peptide bonds, glycosidic bonds, and the like (section 4.4) becomes included in both R (heat loss) and P (production, or the energy of growth materials). Fortunately the amount of energy involved is small, for the bonds mentioned represent less than 1% of the energy of the protein or polysaccharide as a whole. Also, when growth is occurring, heat loss may not have the same relation to oxygen consumption as it does in the fasting animal, because the carbon : oxygen ratio of the new material may not be identical with that of the original respiratory substrates (Wiegert, 1968).

Direct calorimetry is, therefore, the more accurate method of estimating heat loss, but it can involve daunting technical difficulties. Indirect calorimetry brings problems over the conversion of oxygen consumption to heat loss. If both methods are used simultaneously, each serves as a check on the accuracy of the other. The work of Hammen, Lowe, Pamatmat and others, mentioned earlier, involving simultaneous direct and indirect calorimetry, offers a new approach to the study of anaerobic metabolism. Many animals respire anaerobically from time to time, particularly marine invertebrates from temporarily or partially anoxic habitats. Accurate biochemical assays of the various anaerobic end-products (figure 4.5) are difficult to carry out, and calorimetry provides a fresh line of attack. Direct calorimetry measures heat production by both aerobic and anaerobic pathways, whereas indirect calorimetry, based on oxygen consumption measurements, estimates only aerobic heat production. At low ambient oxygen levels direct calorimetry should yield higher values for heat loss than indirect methods, the difference indicating the extent of anaerobic heat production. Figure 5.3 shows results (the means from six specimens) of experiments with a polychaete obtained with the apparatus shown in figure 5.1. At the oxygen concentration equivalent to the 20% air-saturation level, direct and indirect calorimetry indicate mean heat losses of 6.37 and 2.39 joules g^{-1} dry weight hour^{-1} respectively, so at this low ambient oxygen level heat loss from anaerobic processes represents about two-thirds of the total. In fully aerated water, on the other hand, the rate of heat loss was higher and direct and indirect

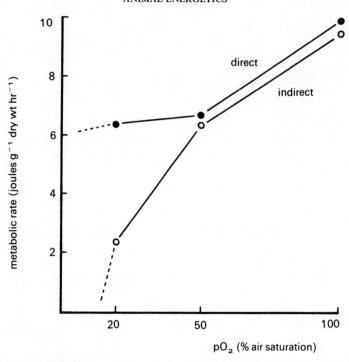

Figure 5.3 Heat loss measured by direct calorimetry, and heat production measured simultaneously by indirect calorimetry, by the polychaete *Neanthes virens* at three ambient oxygen concentrations. (After Lowe, 1978.)

methods produced similar values (9.88 and 9.47 joules g^{-1} hour^{-1} respectively). This is to be expected, and such agreement has usually been found in experiments with other animals. The exception is the work of Hammen (1980) on heat loss by bivalve molluscs, where in aerated water direct calorimetry still gave significantly higher values than indirect—the means from five experiments with the mussel *Mytilus edulis*, for example, being 4.5 and 2.9 joules g^{-1} hour^{-1} respectively. Presumably in this species there can be anaerobic respiration even when environmental oxygen levels are high.

5.2 Muscular activity and the energy cost of locomotion

The harder muscles work, the higher the energy expenditure and the rate of heat production. A variety of terms has come into use to describe the

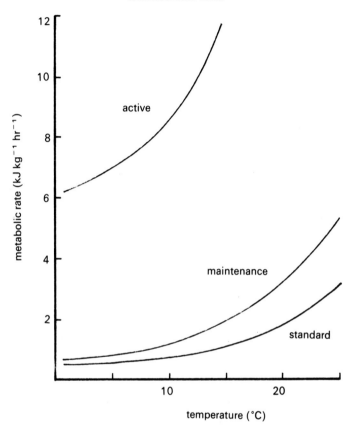

Figure 5.4 Rates of energy expenditure of fingerling sockeye salmon (*Oncorhynchus nerka*) in relation to temperature. The maintenance rate is that shown by fish fed maintenance rations. (After Brett and Groves, 1979.)

various rates of activity and the metabolic rates which accompany them. *Basal metabolism* usually denotes the minimum rate shown by an animal, when it is at complete rest. Only indispensable processes such as blood circulation and the maintenance of muscle tone are taking place. Many animals are never absolutely at rest, however, and the term *standard metabolism* is often used to describe the lowest metabolic rate seen under experimental conditions (figure 5.4). *Active metabolism* denotes the general range of activity that an animal can maintain. An average (70 kg) man expends energy at an hourly rate of about 420 kJ when sitting at rest, over 800 kJ when walking and about 2500 kJ when "jogging". So when a man

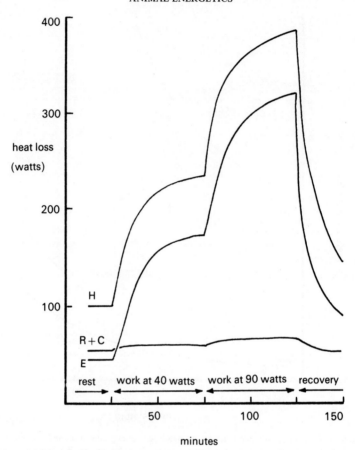

Figure 5.5 The rate of heat loss (in joules sec^{-1}) by a man in a direct calorimeter, when working at the rates shown, at 30°C and 30 % relative humidity. The mean values from eleven men are shown. H—total heat loss; R+C—heat loss by radiation and convection; E—evaporative heat loss. (Simplified from Chappuis *et al.*, 1976).

is working his rate of heat production rises dramatically, but most of the heat is lost by evaporation (in exhalation and through the skin) rather than by convection and radiation (figure 5.5).

Comparing rates of energy expenditure between animals is complicated by the fact that metabolic rate is affected by temperature and by body size. The metabolic rate of a poikilotherm can double if the body temperature rises by 10°C (Q_{10} = 2), but the rise is often less than this, and the animal may acclimatize to some extent (see Somero and Hochachka, 1976). Figure 5.4 shows the effect of temperature on rates of energy expenditure by young sockeye salmon, and such curves are typical of poikilotherms. Large animals produce more heat than small ones overall, but less per unit weight. Thus the metabolic rates of a rat and a cow are about 125 and 3350 kJ day^{-1} respectively, or about 420 and 80 kJ day^{-1} kg^{-1} respectively. Kleiber (1947) plotted heat production per day against body weight, on logarithmic scales, for a wide range of mammals and found that the points fell on a straight line (because R = aW^b, where R is the metabolic rate, W the body mass and a and b are constants). Over the animal kingdom as a whole the same relationship is found, the line having a slope (constant b) of about 0.75 (Hemmingsen, 1960).

Studies of the energy cost of locomotion have concentrated on vertebrates and flying insects. (Information on other animals is sparse and fragmentary, consisting of such snippets as the fact that the motive force in a pseudopodium of an amoeba is about 3×10^{-7} newtons or less, or that a small bivalve mollusc may expend about 2×10^{-3} joules in burying itself in the sand.) Thus most work has been done on the energetics of swimming, flying and running. The rate of energy expenditure can be expressed in terms of the distance covered. Thus if the metabolic rate in joules kg^{-1} hour^{-1} is divided by the speed in km hour^{-1}, the cost of transport will be in joules kg^{-1} km^{-1}. Alternatively, one can consider the rate of expenditure in terms of time, for example in joules sec^{-1} kg^{-1} (equal to watts kg^{-1}).

When a fish increases its swimming speed it increases both the amplitude and frequency of the side-to-side undulations of the body and tail. There are two conspicuous muscle types in fish—red muscle responsible for cruising and white muscle used for short bursts of very fast swimming. The energy cost of swimming is usually estimated by indirect calorimetry, with the fish in a respirometer in a laminar-flow current of known speed. The fish swims into the current and so is effectively almost motionless. As swimming speed increases, the power that the fish must produce increases disproportionately, due to the large drag forces imposed on the fish by the water, and to obtain a straight-line relationship the power has to be plotted on a logarithmic scale (figure 5.6).

The work done during flight must counteract the force of gravity (unlike the case of a neutrally buoyant fish) as well as providing thrust. The energy expended by flying insects varies considerably, but generally lies in the range 0.12 to 0.58 watts g^{-1}. These metabolic rates are among the highest

ANIMAL ENERGETICS

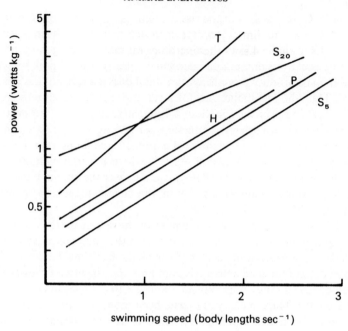

Figure 5.6   The relationship between power input (log scale) and swimming speed for some fishes. T—trout (*Salmo gairdneri*); S_{20} and S_5—sockeye salmon (*Oncorhynchus nerka*) at 20 and at 5°C; H—haddock (*Melanogrammus aeglifinus*); P—pumpkinseed (*Lepomis gibbosus*). (After Goldspink, 1977, from various authors.)

known and often represent about a hundred times the resting rate (Kammer and Heinrich, 1978). In the case of moths, which have a heavy payload and need to warm up before they can fly, the difference between the rate during hovering flight and the resting rate is even greater—about 150-fold according to Bartholomew and Casey (1978), who made the comparison for over 100 species. A 1 g moth, for example, showed an oxygen consumption of about 0.4 cm³ an hour when at rest and about 60 cm³ an hour during hovering flight. Bats show a much smaller difference, the rate of oxygen consumption of *Phyllostomus* in flight being about four times the rate just before flight and about 30 times the rate when totally resting (Thomas and Suthers, 1972).

Most studies on the energetics of flight have concerned birds (see Tucker, 1975a). Energy expenditure is generally estimated by indirect calorimetry. In the apparatus developed by Tucker the bird flies in a wind

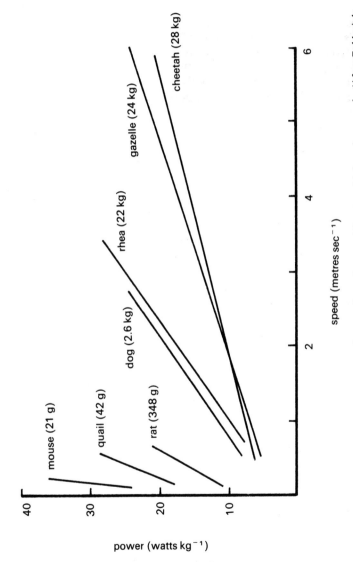

Figure 5.7 The relationship between power input and running speed for some birds and mammals. (After Goldspink, 1977, from various authors, chiefly Taylor.)

tunnel. Pumping air through a mask over the bird's head allows oxygen consumption and carbon dioxide production to be measured. The flight muscles in birds often make up over one-third of the body weight, reflecting the high cost of maintaining flapping flight against drag and gravitational force. Acceleration is achieved by changing the power and configuration of the wing beat rather than its rate. Many birds can glide, thereby saving energy, for as the wings are not beating, muscle activity is isometric rather than isotonic (section 4.3). Hovering, on the other hand, is very costly in terms of energy expenditure, for whereas crows and gulls in flapping flight at speeds of 6 to 12 metres sec^{-1} have power inputs of 50 to 100 watts kg^{-1}, a hovering hummingbird uses about 240 watts kg^{-1} (see Goldspink, 1977). During flapping flight the metabolic rate of a bird is generally about 6 or 7 times the resting level and 10 to 15 times the basal metabolic rate.

Walking and running on land are highly complex activities. Each limb must deliver a power stroke when in contact with the ground and then recover its original position in order to repeat the process. The centre of gravity may rise and fall with the stride, wasting energy, and the animal must keep its balance. Each limb is constantly accelerating and decelerating, and the body is under some vertical compression as it is on a rigid substratum while subject to the force of gravity. Alexander (1977) has considered the biomechanics of walking in detail. The energy cost of walking or running is generally easier to estimate than that of swimming or flying. The animal can often be run on a treadmill, so the speed of locomotion can readily be measured and controlled. A running animal shows a straight-line relationship between power input and speed of movement, although the position and slope of the line varies widely according to the species (figure 5.7). Most of the energy is apparently spent in tensioning the tendons that serve to store energy, and in doing other internal work. Comparatively little is used in accelerating and decelerating the limbs (Goldspink, 1977).

To summarize, a fish when swimming generally uses less than 5 watts kg^{-1}, a running mammal in the region of 5 to 35 watts kg^{-1}, and a bird in flapping flight about 50 to 150 watts kg^{-1}. The low cost for the fish reflects the fact that although it has high drag forces to overcome, gravity presents no problem. The high cost for the bird results from having to keep itself airborne against the force of gravity. On the other hand, birds generally fly at higher speeds than those attained by running animals, and so they cover a given distance for a smaller expenditure of energy than do terrestrial animals. This is evident from figure 5.8, which also shows that

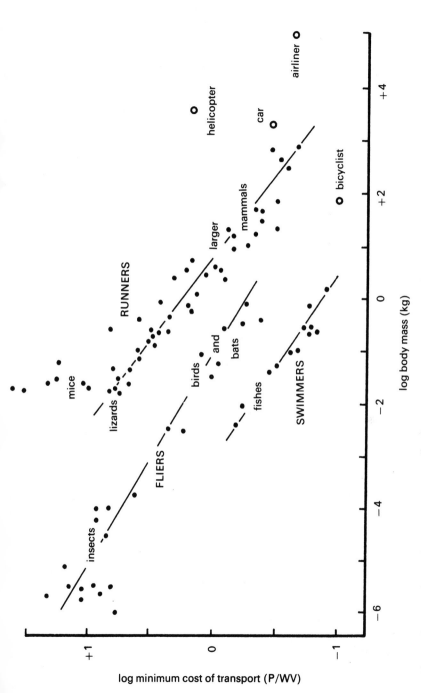

Figure 5.8 The relationship between cost of transport and body mass. Some of the central values among the runners are for birds. The cost of transport is expressed as the power input in watts (P) divided by the weight (W = mass in kg × acceleration due to gravity) and the speed of movement in metres sec^{-1} (V). (After Tucker, 1975b, based on data from many authors.)

in all three groups (swimmers, fliers and runners) larger animals have a lower cost of transport per unit weight than smaller ones.

5.3 Other sources of heat

Although active muscles may be responsible for most of the heat an animal produces, other organs and tissues inevitably produce heat in the course of their metabolism. It is difficult to make comparisons between the heat production of various organs, but Woledge (1980) has reviewed the subject and reports figures of about 4 to 8 watts kg^{-1} for brain tissue and 30 watts kg^{-1} for kidney tissue. Although the liver, a large and metabolically active organ, must have a significant rate of heat production, this has not yet been measured in absolute units (Woledge, 1980).

The tissue that has attracted most attention is the brown fat of mammals, which is generally most conspicuous in the young. It accounts for about 5% of the body weight of a newborn rabbit, where it is concentrated in the neck and thorax. The breakdown of triglycerides into fatty acids and resynthesis of the latter back into triglycerides, which occurs in brown fat, is futile except that it produces heat which helps the young mammal compensate for having very little fur to check heat loss. Brown fat is very active metabolically and its high concentration of cytochrome-containing mitochondria (section 4.2) gives it the brownish colour from which the name derives. Much heat is produced by the mitochondria of brown fat and heat, rather than ATP, is their main product. (Oxidative phosphorylation is probably uncoupled from electron transport.) Heat production by brown fat is also important to hibernating mammals during arousal in the spring. Until recently it was thought that other adult mammals had little use for brown fat, but it now seems possible that lean rats and humans are able to eat a lot without becoming obese because their brown fat "burns off" the energy as heat.

Heat production by brown fat is one aspect of what is known as *non-shivering thermogenesis*. Specific dynamic action (section 3.4) is another. The production of heat "on purpose", rather than as an inevitable and "unwanted" consequence of the inefficiency of metabolism primarily concerned with ATP production and use, is a relatively new concept. In cases where heat is useful in its own right, even inefficient or apparently futile metabolic pathways can have an important role to play. The uncoupling of mitochondrial oxidative phosphorylation, and the transformation of triglycerides to fatty acids and back again, have been mentioned above. These and other thermogenic pathways have been discussed by Stirling and Stock (1973).

5.4   Thermoregulation

Birds and mammals are homeotherms (endotherms) with a high metabolic rate and the ability to regulate their body temperature within very narrow limits. When the ambient temperature is too low they increase the rate of heat production and when it is too high they increase the rate of heat loss. Other animals are known as poikilotherms (or ectotherms) and their body temperature tends to change with that of the environment. The distinction between the two groups is becoming less clear, however, as the following examples show. The best temperature for the growth and activity of sockeye salmon is 15°C, and given a choice these fish will select water of this temperature. Several fish, amphibians and lizards, given a choice of two non-optimal temperatures, move back and forth between them in such a way that they maintain their preferred body temperature (Heller *et al.*, 1978). The bluefin tuna has a specialized vascular system that helps to conserve metabolic heat, so successfully that its maximum muscle temperature is about 27°C when the water is at 10°C, and 32°C when the water is at 30°C (Carey, 1973). Some lizards and turtles can keep their body temperatures below high ambient levels by panting, and bees and termites control the temperature of their colonies.

As the ambient temperature drops, heat will be lost from the body more quickly and the body temperature will fall unless some action is taken. In such circumstances a mammal can fluff out its fur, or a bird its feathers, thus checking heat loss. Constricting the superficial blood vessels has the same effect. At the same time heat production by a homeotherm can be increased by shivering. Conversely, if the temperature rises, the rate of heat loss can be increased by dilation of the superficial vessels and by increasing evaporative heat loss. Mammals such as man and the horse, which have many sweat glands, can achieve this chiefly by the evaporation of sweat, but the dog must rely on panting. At a skin temperature of 37°C the latent heat of vaporization will be about $2.41 \, \text{kJ} \, \text{g}^{-1}$ of sweat. An average man has over two million sweat glands and heat loss through sweating can be considerable. (In 1798 a man stayed for 45 minutes in a dry room heated to 126°C.) In addition to these physiological adaptations a homeotherm faced with heat or cold stress may seek out a cooler or warmer micro-climate, or create one for itself.

The thermostat of a mammal is in the hypothalamus. Hammel (see Heller *et al.*, 1978) has implanted thermodes (minute tubes) around the hypothalamus of dogs and perfused these with water of a specific temperature. When the hypothalamus was warmed in this way panting was induced, above a certain threshold temperature. Similarly, when the hypothalamus was chilled, shivering was induced below a threshold temperature (figure 5.9A). Clearly the hypothalamus has both warmth and cold receptors, but such receptors are also located in the skin. (One sweats

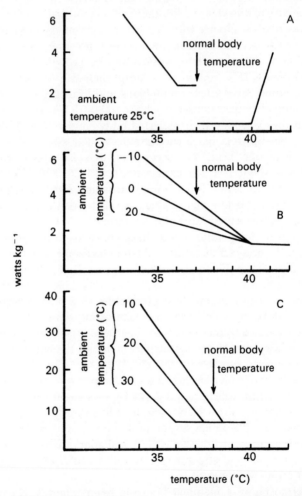

Figure 5.9 (A) Heat production (left) and evaporative heat loss (right) by a dog when the hypothalamic temperature (horizontal axis) is altered by means of thermodes. (B) The relationship between heat production and central temperature in the harbour seal, at three ambient temperatures. (C) As (B), for the kangaroo rat. (After Heller *et al.*, 1978.)

almost at once on entering a hot environment, before the hypothalamic temperature can have risen to the threshold temperature.) These superficial sensors pass information about environmental temperature to the hypo- thalamus so that changes in central temperature can be anticipated and avoided. Thus the responses of skin receptors to cold can result in the hypothalamus initiating shivering in time to avoid a drop in central temperature. The effect of a fall in ambient temperature on the hypo- thalamic thermostat varies with the species. It can either increase the sensitivity, causing a greater rate of heat production (as in the harbour seal, figure 5.9B), or it can increase the threshold temperature (as in the kangaroo rat, figure 5.9C). Thus environmental temperature changes are sensed by skin receptors and the appropriate responses are initiated and controlled by the hypothalamus. During physical exertion sweating is induced by the hypothalamus in response to the rise in central temperature, but it is uncertain whether the receptors involved are in the hypothalamus or other areas, such as the spinal cord. In the dog there is input to the hypothalamus from receptors in the muscles and joints, so that panting starts soon enough to prevent a rise in central temperature (Heller *et al.*, 1978).

Homeotherms in very hot or very cold climates have a particularly severe problem. Those of hot deserts seek a cooler microclimate, such as a burrow, during the hottest part of the day. They only use evaporative heat loss as a way of controlling their temperature in emergencies, because water is in short supply. Arctic mammals are well insulated by fur or fat, except at the extremities. The nose and paws can be hidden away by "balling up", and they serve as a valuable channel for losing body heat after exertion. The feet of birds and mammals of cold climates are adapted to withstand very low tissue temperatures, and heat loss from them is often reduced by a counter-current system in the limbs whereby heat can pass from arteries to nearby veins and consequently not be lost from the extremities.

SUMMARY

1. Energy lost as heat, represented by R in the energy budget equation, is an indication of metabolic rate, the rate of ATP turnover. R can be estimated by either direct or indirect calorimetry.

2. In direct calorimetry the heat lost by an animal in an insulated chamber is measured by thermistors or thermocouples. The technique can be used with both air-breathing and aquatic animals but to be accurate it requires rather elaborate technology.

3. In indirect calorimetry oxygen consumption can be measured and multiplied by an oxycalorific coefficient (Q_{ox}) to estimate heat loss. The Q_{ox} varies with the respiratory substrate: for carbohydrate, for example, it is 14.76 joules mg^{-1} oxygen consumed. More reliable estimates of R are obtained in cases where equations involving the rates of carbon dioxide production and nitrogenous excretion (in addition to oxygen consumption) can be applied.

4. Direct calorimetry takes account of both aerobic and anaerobic respiration, whereas indirect calorimetry based on oxygen consumption measurement estimates only aerobic heat production. Thus simultaneous use of both techniques can indicate the extent of anaerobic metabolism.

5. Metabolic rate varies with ambient temperature, with body size, and with the activity of the animal. The rate of energy expenditure for a swimming fish is about 5 watts kg^{-1}, for a running mammal about 5 to 35 watts kg^{-1} and for a bird in flapping flight about 50 to 150 watts kg^{-1}. The bird must work to keep itself airborne against the force of gravity. On the other hand birds generally fly faster than terrestrial animals run, and so cover a given distance for a smaller expenditure of energy.

6. Although muscles may be responsible for most of the heat an animal produces, other organs and tissues also make a contribution. Brown fat is particularly active, and its otherwise futile biochemical pathways produce heat of great value to the poorly-insulated young mammal and to hibernating mammals during arousal.

7. Birds and mammals (homeotherms) regulate their body temperature within very narrow limits; numerous other animals can also control their temperature, but to a limited extent. When the ambient temperature falls a homeotherm increases heat production and reduces heat loss by various means. Increased heat production resulting from activity requires a homeotherm to increase the rate of heat loss, for example by the evaporation of sweat or by panting. The thermostat of a mammal is in the hypothalamus.

CHAPTER SIX

PRODUCTION

Production is mainly channelled into increase in the animal's biomass (and therefore energy content) or into the formation of gametes and embryos. So the majority of production, P, is usually associated with growth (Pg) or with reproduction (Pr) and this chapter is concerned with these processes. Animals make other materials, however. The silk of web-spinning spiders and silkworms, the cocoons of earthworms and caterpillars, and the wax produced by bees also form part of production. In addition, arthropods moult at intervals and the energy in the shed "skins" is included in P when erecting energy budgets. Such materials usually represent only a small part of input energy (C) but Teal (1957) found that up to 70% of C can be used for mucus production by flatworms, where it aids their locomotion.

6.1 The energy cost of growth

Animals ingest macromolecules and break them down by digestion into monomers. These are assimilated from the gut. Some are used in respiration but some are assembled into the polymers which comprise animal production. As the new tissue has a higher energy content than the constituent monomers, energy is expended in its formation. The energy cost of these processes depends on the energy requirement of the biochemical reactions involved. At the same time, the animal has to expend energy to maintain its existing body tissues for there is a continual breakdown and reassembly of cell constituents.

The efficiency of the conversion of energy consumed (C) into production (P) is considered in chapter 8. Here we are concerned with the energetics of the biochemical processes involved in growth. Calow (1977) has calculated the energy required to build polymers from monomers to produce animal tissue. His energy budget for production is given in

71

figure 6.1. An energy input from ATP of 860 joules is required to synthesize 1 g of biomass, on average, and the cost of producing the necessary ATP is 1323 joules. Only 241 of these are actually incorporated into the polymers, however, the remaining 1082 being lost as heat during ATP production (463 J) and polymer synthesis (619 J).

The energy content of the gram of biomass is 23 100 J, of which 241 J is derived from ATP and the remainder from the energy already in the monomers. The total cost of the product is therefore 23 100 J plus 1323 J (the energy associated with ATP), or 24 423 J. The net conversion efficiency is

$$\left(1 - \frac{\text{cost of production}}{\text{total amount used}}\right) \times \frac{100}{1} = \left(1 - \frac{1082}{24\,423}\right) \times \frac{100}{1} = 96\,\%$$

Thus biological systems can be extremely efficient in the production of new tissue from raw materials. The efficiency of 96% is the highest possible, because there is a cost in assembling the super-molecular organization of biological systems. This is probably negligible, however, because the heat of combustion of the organized cellular organelles and the cells themselves is very nearly equal to the combined heats of combustion of the constituent macromolecules. Calow (1977) did not explicitly include the cost of digestion or the effect of differences in the energy content of absorbed monomers and new polymers when he calculated the energy cost of production.

6.2   Control of animal growth

The growth rate of an animal (the speed with which it incorporates the energy content of food into biomass) and the final size it attains are likely to be important features of its "fitness", determining whether or not it lives to reproduce and pass on its genes to the next generation. Many environmental factors can influence growth, such as the availability and quality of food and the temperature, but we can expect natural selection to have built some sort of regulatory mechanism into the animal (Calow, 1976a). Many people have fitted mathematical equations to growth curves but these often only describe or summarize the data and contribute little to understanding what dictates the rate. In 1938 von Bertalanffy made one of the first attempts to explain animal growth rates. He considered that growth rate was a function of two processes, synthesis and degradation (anabolism and catabolism), and that these were dependent on body size.

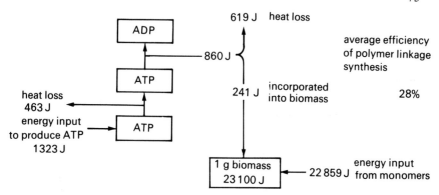

Figure 6.1 The energetics of biomass production. (Derived from Calow, 1977.)

The larger the organism the greater were the two processes, but both could be non-linear functions of weight. His growth equation is:

$$\frac{dW}{dt} = hW^n - kW^m \tag{6.1}$$

where the change in weight W with time t is related to size by the powers n and m, and h and k are constants associated with anabolism and catabolism respectively. To evaluate n, von Bertalanffy assumed that anabolism would be proportional to the absorptive surface of the organism, as it relies upon raw materials being taken up. Catabolism, involving m, should be a function of mass, as it is associated with metabolic rate. Thus anabolism could be related to the square of the linear dimension and catabolism to the cube.

This model attempts to explain growth, rather than simply describing the pattern of size change with time, and its parameters can be determined empirically and independently. It has been used extensively for estimating the growth of field populations, by Beverton and Holt (1957) for instance, but has also been criticized as a gross over-simplification of a complex process. Calow (1976b) criticizes the model chiefly because it implies that the growth process is entirely passive, and unable to make any immediate compensation for changes of food supply and temperature. He feels there is ample evidence for the involvement of various regulatory mechanisms concerned with feeding and that this should lead us to consider growth models which include positive feedback.

An alternative to von Bertalanffy's equation is the growth model developed by Hubbell (1971) and summarized by Calow (1976b). This

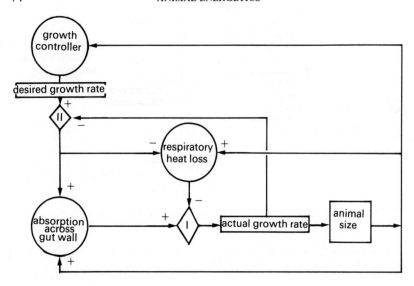

Figure 6.2   The flow diagram of Hubbell's growth model. (Adapted from a simplification by Calow, 1976*b*).

model is depicted in figure 6.2. It takes into account the active regulation of growth by feedback control on anabolism and catabolism (so the essence of the model is similar to that of von Bertalanffy) with the growth rate depending on "absorption across the gut wall" and on "respiratory heat loss". Absorption has a positive value and respiration a negative one. The difference between these signals, calculated at comparator I, gives current actual growth rate, which upon integration gives current size. Current size has a positive influence on both anabolism and catabolism and as we have already seen these are size-dependent processes. Hubbell's model extends the concept of growth by the inclusion of a growth controller which generates a "desired growth rate". The growth controller is in turn influenced by animal size, because in young animals growth rate is usually size-dependent. Comparator II evaluates growth performance by subtracting "actual" from "desired" growth rate and generates a feedback to respiratory heat loss (negative) and absorption (positive). Hence an animal which is starving will have reduced respiration but an increased efficiency of digestion. All three sub-systems are subject to environmental disturbances and all relationships may operate in a non-linear fashion.

This model has been built up from general biological observations and intuition, but do animals actually have growth controllers, feedback pathways and comparators? Animal growth involves cells, tissues, organs and the organism as a whole, and each level has specific control mechanisms, ranging from enzyme regulation in cells to specific inhibitory and stimulatory secretions acting on the organs (Needham, 1964). Some mammals and birds appear to possess a serum factor which depresses overall growth and which increases in potency (or concentration) with age. This may be a steroid from the adrenal cortex which antagonizes the pituitary growth-stimulating hormone. Such a mechanism would act as an overall growth controller. An example of an "active" feedback mechanism, which could form part of Hubbell's model, is the regulation of energy intake by the hypothalamus (section 3.2). The control of respiratory heat loss, however, can be "passive", for respiratory rate may be proportional to the level of glucose circulating in the blood. The growth model may, then, have both "active" and "passive" regulation mechanisms, but the way the model behaves when it is constructed and operated suggests that "active" rather than "passive" control may underlie growth of an organism. Hence if growth is regulated in such a way, animals make positive "decisions" which allow them to grow at an optimal rate under a given set of environmental conditions.

6.3 The partitioning of energy between growth and reproduction

The development of reproductive organs is usually considered part of the normal growth process, and the energy content of the ovaries or testes of a mammal may be quite a small proportion of the overall biomass at puberty. A parthenogenetically reproducing insect, in its final larval instar, may have most of the energy content of its ovarioles packaged as embryos, soon to be shed. Biomass synthesized within a mother but eventually ejected as offspring is usually accounted for separately from animal growth. The balance between growth and reproduction is important in the long-term survival of the species.

Hubbell's growth model can be modified to account for the reproductive portion of production (figure 6.3) by representing reproduction as a negative input signal from Comparator III to the body's potential energy store (Comparator I). Comparator III—the reproduction input generator—receives information from two sources. There is a positive input related to body size and in Hubbell's model this is a linear relationship between animal size and reproduction—the bigger the animal the greater its reproductive capacity. This is clearly a simplification

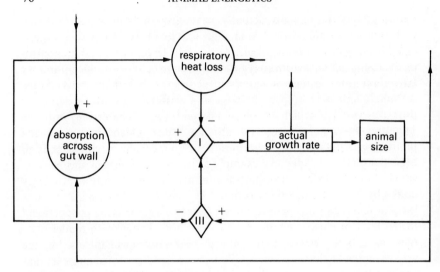

Figure 6.3 Part of Hubbell's growth model modified to include reproduction. (Adapted from a simplification by Calow, 1976*b*).

because there is often a body size below which no reproduction is possible—see Taylor (1975) for example. There is a negative input, from the feedback, on desired animal growth generated by Comparator II. Here Hubbell is suggesting that if the requirement for parental somatic growth is high then there is an increased negative input into Comparator III and less energy is channelled into reproduction. On some occasions it is preferable for an animal to grow rather than reproduce and this overall pattern is determined by the growth controller (figure 6.2) acting on Comparator II.

If a parent reproduces on one occasion and then dies, as in many insects and polychaetes, (this is known as *semelparity*), then the number of progeny born will be the maximum possible for a given animal size. Once adult size has been reached there should be no need for the negative feedback from "desired growth". Some animals, however, reproduce in successive breeding seasons (*iteroparity*) and this allows for developments such as parental care and transmission of information through behaviour. In these cases we need to include the negative feedback associated with "desired growth", because reproduction should only rise to a level which does not hamper the future well-being (and hence total reproductive potential) of the parent. There are numerous cases where starvation of iteroparous parents leads to reduced fecundity—short-term reproduction is sacrificed for long-term survival of the parent.

Few reviews concerned with energetics have considered the magnitude of the different fates of production, but Conover (1978) assembled about twenty energy budgets for aquatic invertebrates and then calculated the portion of assimilated energy channelled into growth (Pg) and reproduction (Pr). He found that, on average, 20% of assimilated energy was used for growth and 15% for reproduction. There were great discrepancies in the budgets he considered, apparently dependent on the method of reproduction and the time over which the budget was calculated. The production of a few large yolky eggs or numerous small poorly-provisioned eggs was important and often the energy budget spanned only part of the growth and reproductive cycle. When data were compiled for intertidal or subtidal invertebrate populations, where the energy budget had been calculated for the whole year, on average 19% of assimilated energy was channelled into growth and only 4.3% into reproduction. Maintenance costs for these populations are often very high, presumably a consequence of winter survival with little feeding, and so there is little energy remaining for reproduction.

Examples of production partitioning among terrestrial invertebrates also show great variability. In the black bean aphid *Aphis fabae* (a parthenogenetic, viviparous insect) the energy content of Pr is not only derived from parental energy intake during reproductive life but also from catabolism of the parent biomass during this period (Llewellyn and Qureshi, 1978). This insect feeds by piercing phloem sieve tubes and extracting the plant sap, and when reared on mature plant leaves or on leaves and stems of young plants, 5% of the energy content of reproduction originated from catabolism of the parent biomass. When aphids were reared on the stems of mature plants, however, up to 20% of the energy content of reproduction was derived from the mother's biomass. When the energy budget for the whole life cycle was calculated, for all feeding sites, larval growth accounted for 18% of total assimilated energy and reproduction for 70%. Reproduction accounted for 78% of total production.

McNeill (1971) constructed energy budgets for an insect population with a very different reproductive style to that of aphids. Mirid bugs (*Leptopterna dolabrata*) feed on the cell contents of leaves and ovules of various grasses, especially *Holcus mollis* and *H. ianatus*. These insects produce only one generation a year compared to the 8–10 day generation time of *Aphis fabae*. All the production attributed to the adult insects was judged to be eggs and sperm (termed the reproductive effort) although not

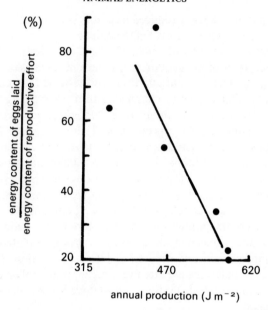

Figure 6.4 The efficiency of egg production of *Leptopterna dolabrata* related to annual production. (After McNeill, 1971.)

all eggs survived long enough to be laid. Reproductive efficiency (the energy content of eggs produced divided by the energy content of reproductive effort, expressed as a percentage) was related to annual production (figure 6.4) and we see that as total production increases the energy successfully channelled into reproduction declines. This has the effect of reducing the fluctuation in production from year to year and hence smooths out population number fluctuations. Reproduction accounted for 22% of production, a significantly lower proportion than seen with the aphid populations.

6.4 Strategies of reproduction

We have considered how much of P is directed into Pr but what determines the pattern of reproduction? Where resources are effectively unlimited the fitness of a population—how it will succeed both in the short term and in the longer term—is raised by increasing the number of progeny. This may be achieved by increasing the birth rate and by reducing the time between breeding periods.

We would expect small, fecund organisms to evolve under such conditions. If resources become depleted, however, the capacity of the population for increase will become impaired, and as the number of organisms in the population declines to the maximum that the environment can support, reproduction will fall to a level which allows only for replacement, if that. Under these conditions "fitness" is the success with which populations procure sufficient of the limited resources to enable survival. Offspring produced by such animals must be competitive and are likely to be larger and therefore fewer in number. A fall in reproduction is necessary since the diversion of too great a proportion of limited resources into it is likely to put the parents, and hence their reproductive potential, at risk. These two reproductive strategies or traits have arisen because of the differing amounts of environmental resources available. They have been referred to as "r" and "K" selection by MacArthur and Wilson (1967), where "r" refers to the strategy involving many small and fecund organisms, and "K" to the opposite condition. In reality most habitats offer intermediate conditions so that animals will tend either to the "r" end of the spectrum or the "K" end.

A high reproductive output coupled with repeated breeding (iteroparity) provides for immediate population increase, effective exploitation of environmental resources and insurance against mass loss of progeny. This seems to occur only rarely (e.g. endoparasites) and usually there is an inverse relationship between reproductive output and repeated breeding (Calow, 1978). If energy is used for reproduction when it is required for other metabolic processes there will presumably be a serious risk to the parent and its reproductive potential. What is important, in terms of fitness of the population, is not the reproductive performance of an organism in any one period but its lifetime reproductive performance. This has two distinct components—present reproductive output and possible reproductive performance in the future. Present reproductive output can only be increased to advantage if the gains made are greater than the reduction in potential reproductive performance. When the environment favours the evolution of "r" species (most resources going into reproduction) future reproductive potential is less important. Mortality tends to be density-independent under such environmental conditions, and this will further reduce future reproductive performance because mortality in such circumstances is often age-independent, with offspring and parents equally likely to die. This suggests that iteroparity (repeated breeding) will be a common feature of "K" selected organisms. Reproductive output will presumably be greatest when mortality is erratic and least when density-dependent mortality (usually associated with environmental conditions favouring "K" selection) predominates. The "r" selected species are the colonizers or opportunists of the ecosystem, responding to and taking advantage of a variable environment. "K" selected species, on the other hand, adapt to take advantage of limited resources and are the long-term occupants of stable ecosystems. Stearns (1977) questions the usefulness of the "r" and "K" concept because he finds it difficult to make unambiguous predictions from it for natural populations, even where the population dynamics have been precisely defined. It is easy to see how this concept of reproductive strategy might be modified. Low replacement and high juvenile mortality may occur, for instance, when most of the suitable sites for establishing a population have been filled, or when they were few and far between to start with. This is characteristic of "K" selected situations but might also occur in colonizing situations which in other respects might encourage "r" selected species.

The strategy of reproduction, then, is assumed to be adapted, through evolution, to cope with particular environmental situations. It is likely that during the life of any individual animal environmental conditions may change, however, and if so to what extent can individual reproductive performance be altered? In section 6.2 we suggested that organisms partition energy consumed (C) into that required for metabolic processes (R) and that required or remaining for growth and reproduction (P). In many species, for animals of a given size and age, a fixed amount of energy may be required for metabolism and failure to supply it may be detrimental to the animal. If R remains constant and C is reduced, the ratio Pr/C must be reduced. When comparing the reproductive performances of individuals and populations at different levels of food ration the ratio Pr/C is rather crude. Calow (1978) has suggested that the expression

$$1 - \frac{(C - Pr)}{R^*} \qquad (6.2)$$

is more appropriate, where R^* is the energy used in metabolism under optimal conditions. This is an expression of reproductive effort. If it remains constant (figure 6.5a) then the ratio Pr/C must alter when food rations decline (figure 6.5b). If organisms partition energy so that Pr/C falls below the line in figure 6.5b then they are exhibiting "reproductive restraint" in times of food shortage, since reproductive effort falls proportionately more than the reduction in ration. Species which allocate energy so that Pr/C is above the line are showing "reproductive recklessness" because they reproduce at the expense of R, that is Pr/C remains steady

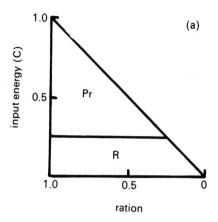

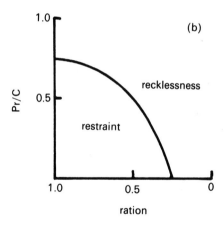

Figure 6.5 Reproductive effort and energy partitioning of animals. See text for explanation. (After Calow, 1978.)

(or rises) as the food ration is reduced. These animals must have short-term metabolic flexibility, being able to reduce the amount of C which is apportioned to R when food rations are low. "Recklessness" is an appropriate strategy when the progeny have an equal or better chance of surviving the food shortage than their parents. "Restraint" is more appropriate when parents are better able to withstand food shortage than their progeny.

6.5 Growth and reproduction rates

Whilst many authors recognize that growth may encompass repair and maintenance (Needham, 1964), production (P) is considered to include only increase in biomass or a change in body constituents resulting in an increased energy equivalent. As growth is essentially quantitative its most important property is its rate, which varies enormously. The blue whale achieves a weight of over 100000 kg in about 5 years, a rate of about 56 kg day^{-1}. Man takes about 20 years to reach a weight of 70 kg, an average of about 1 g day^{-1}. The black bean aphid gains 0.8 mg in 7 days, or about 0.1 mg a day. What results from growth is itself capable of growth, and if growth rates were constant the graph of mass against time would be exponential. Early in life the proportion of cells that grow and multiply actively is close to 100 % but the proportion falls later. There are several ways of dealing with data of weight against time, which reflect animal growth. A growth curve results if we simply plot the weight at each interval of time (figure 6.6a) but a common and useful alternative is to calculate the *instantaneous* (or *specific*) growth rate. This expresses the daily (or weekly or yearly) increment per unit of existing biomass and is given as a percentage by the equation

$$IG = \frac{\log_{10} YT - \log_{10} Yt}{T - t} \times 2.3026 \times 100 \tag{6.3}$$

where t is the time at the beginning of the observation, T that at the end, YT is the weight at time T and Yt that at time t. 2.3026 converts $\log_e$ into the more familiar $\log_{10}$. Such a plot of growth rate for the data of figure 6.6a is seen in figure 6.6b. It assumes that all the tissue of the body is growing at the same rate and that all newly-formed tissue starts to grow at that rate as soon as it is formed. The instantaneous growth rate is a good description of how an organism grows but the time intervals should be as short as possible when size is changing rapidly. Figure 6.6b has a negative exponent, with an initially steep decline becoming progressively less steep and tending asymptotically to a constant low value. The curve for growth rate of figure 6.6a is exponential in its early stages but reaches an upper size limit.

Animals usually produce numerous progeny and this is our concern when quantifying Pr. Animal populations are usually contained within relatively narrow limits and so considerable juvenile mortality must occur. For example, of about 60 million oyster larvae produced by one adult, only about 200 metamorphosed oysters survive (Cohen, 1977). The

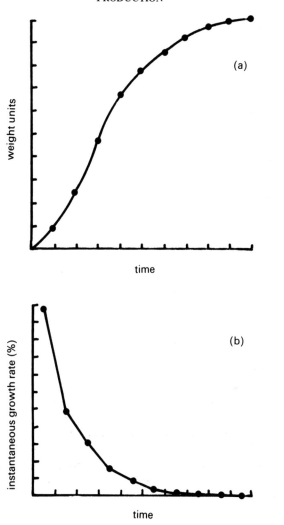

Figure 6.6 Animal growth curves. (a) weight of animal at each interval of time. (b) the instantaneous growth rate for the same data. (After Open University S 323.)

numbers of progeny produced by females of a variety of species are given in table 6.1. The strategy of reproduction adopted, as we have seen in section 6.4, depends upon the relative chances of survival of large and small eggs or offspring. As conditions for offspring survival become

Table 6.1 Numbers of offspring produced by breeding female animals during their lifetime (after Cohen, 1977)

	Number brood^{-1} season^{-1}	Estimated total production
Oryctolagus cuniculus (rabbit)	8	25
Sturnus vulgaris (starling)	4.8	16
Parus major (great tit)	10.3	30
Anguis fragilis (slow-worm)	7–19	40
Gadus morhua (cod)	2–20 million	50 million
Salmo salar (salmon)	7–10 000	7000
Scyliorhinus canicula (dogfish)	8–20	40
Crassostraea virginiana (oyster)	1 million	3 million
Apis mellifera (honey bee)	120 000	0.5 million
Anomma wilverthi (African river ant)	40 million	240 million
Musca domestica (housefly)	100	100
Aphis fabae (black bean aphid)	70	70

tougher, eggs may become larger. Gastropods living in freshwater habitats produce between 10 and 10^3 eggs per individual in a lifetime whereas marine species produce about 10^6 eggs per individual in a lifetime. As gastropods have invaded freshwater habitats, free-living larval stages have been retained in an envelope together with a supply of food. This requires more energy input per egg and there may be a reduction in number (Russell-Hunter, 1970).

Many environmental factors affect the rates of growth and reproduction, and growth rate often declines as an animal ages. Some animals grow at a steadily declining rate whereas others grow sporadically. Where the environment alters seasonally, cyclical changes in growth can occur in response to changes in temperature and food supply, for instance. In addition some fish, for example, show endogenous rhythms, growth slowing or ceasing as the breeding season approaches and resuming afterwards.

6.6   The prediction of production from respiration and biomass data

In 1966 Engelmann related animal production to annual respiration for some terrestrial animal populations and commented that such a relationship might be a useful ecological tool for estimating production, because respiration is relatively easy to measure. McNeill and Lawton (1970) re-examined this relationship in the light of a large number of further studies and taking into account aquatic animals. If assimilation (A) is

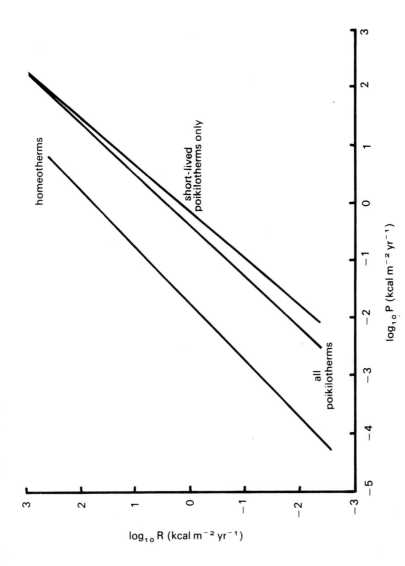

Figure 6.7 The relationship between annual respiration and annual production. (After McNeill and Lawton, 1970.)

taken to be the sum of respiration (R) and production (P), a ratio of P to R allows calculation of P/A (see section 8.3). The equations relating P to R derived by McNeill and Lawton (1970) are increasingly being used to complete energy budgets from a knowledge of either production or respiration. They were arrived at from logarithmic plots of R against P (figure 6.7). Direct and independent field measurements of P and R are impossible because both are calculated from estimates of population numbers or biomass. Therefore P and R are not statistically independent variables and normal regression techniques should strictly speaking not be used, but McNeill and Lawton felt the relationships were sufficiently important to justify doing so. They used data from 53 sources, and obtained the following regression equations:

all heterotherms
(poikilotherms)

$$\log R = 1.0733 \log P + 0.3757 \qquad (6.4)$$

$$or \quad \log P = 0.8233 \log R - 0.2367 \qquad (6.5)$$

short-lived
heterotherms only

$$\log R = 1.1740 \log P + 0.1352 \qquad (6.6)$$

$$or \quad \log P = 0.8262 \log R - 0.0948 \qquad (6.7)$$

homeotherms

$$\log R = 0.9812 \log P + 1.7418 \qquad (6.8)$$

$$or \quad \log P = 1.0137 \log R - 1.7761 \qquad (6.9)$$

The slopes of the· lines in equations 6.4, 6.8 and 6.9 do not differ significantly from unity but those for the other three do. Calculated 95 % confidence limits for these regressions indicate a considerable scatter of data around the fitted lines. It is clear from figure 6.7 that energy expended on respiration at any level of production tends to be higher in long-lived heterotherms, where a standing crop of older individuals is maintained from year to year. Miller *et al.* (1971) accumulated the confidence limits through a series of calculations, starting from individual respiration measurements and ending with a final annual production estimate, for a population of lobsters in eastern Canada. The final upper limit was 446 % greater than the mean! But the estimated production using the production/ respiration relationship was $223 \, kJ \, m^{-2} \, yr^{-1}$ and compares well with the measured production of $209 \, kJ \, m^{-2} \, yr^{-1}$ (Miller and Mann, 1973).

Production can also be related to biomass, using the expression:

$$P = IG\bar{B} \qquad (6.10)$$

where $\bar{B}$ is the average biomass during a time interval t and IG is the instantaneous growth rate expressed as a weight. This relationship only

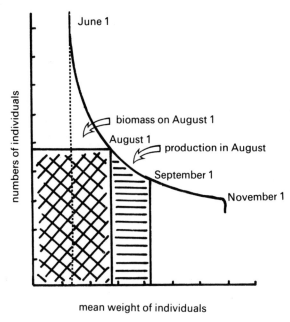

mean weight of individuals

Figure 6.8   Allen's graphical method for estimating production and biomass. (After Open University S 323.)

holds if there have been no deaths, emigration or immigration during the time interval. This is unrealistic for a field population, but we shall see in chapter 9 how equation 6.10 may be adapted and used to estimate P for natural populations. It is satisfactory over short periods, or when IG and mortality are both small. Allen (1951) devised a simple graphical method for estimating production and biomass when he plotted the number of individuals at successive times against the mean weight of individuals at those times. If the area under the curve for a particular time period equals production and can be estimated by direct counting, biomass can be deduced from the graph (figure 6.8) by constructing an appropriate rectangle.

SUMMARY

1. The energy content of a gram of biomass is 23 100 J, of which 241 J are derived from ATP and the remainder from the energy already in the monomers. 1323 J are required to manufacture the ATP and so the total cost of the product is 24 423 J, which yields a net conversion

efficiency of 96 %. Thus biological systems can be extremely efficient in the production of new tissues from raw materials.

2. Growth models developed by Hubbell are presented as an alternative to Von Bertalanffy's equation. Hubbell took into account the active regulation of growth by feedback control on anabolism and catabolism, which yielded current "actual growth rate". Upon integration this gives current size. A growth controller generates a "desired growth rate" and a comparator evaluates growth performance by subtracting "actual" from "desired" growth rate.

3. Hubbell's growth models can be modified to account for different patterns of animal reproduction. Semelparity (rapid reproduction followed by parental death), characteristic of "r" selected animals, and iteroparity (reproduction by adults in successive breeding seasons), characteristic of "K" selected species, are accounted for. These reproductive traits have arisen because of differing environmental resources available. In reality most habitats offer intermediate conditions.

4. Animal growth is best represented by the instantaneous (or specific) growth rate, which assumes that all body tissue is growing at the same rate and that all newly-formed tissue starts to grow at that rate as soon as it is formed. Instantaneous growth rates are a good description of how animals grow, providing the time intervals are short.

5. There have been several attempts to equate annual animal production with annual animal respiration. This is so that the value of one parameter may be derived from a measure of the other. Such relationships can be used to complete production estimates for use in animal energy budgets.

CHAPTER SEVEN

ENERGY OF WASTE PRODUCTS

Two categories of energy-rich waste products are eliminated from animals. Faeces are formed from food material which has not been digested, or has been digested but not assimilated. Urine contains the nitrogenous products of catabolism, of which ammonia, urea and uric acid are the commonest. Faeces and urine have separate origins and distinctive chemical compositions. For an energy budget they are measured as separate energy pathways, F and U, where possible but in some animals urine is voided into the alimentary canal and mixes with faeces before elimination. In such cases, insects for example, U can be distinguished from F by determining the amount of nitrogenous waste chemically and multiplying by the energy value per unit weight. Alternatively, U can be calculated from R, if the amount of oxygen used in respiring protein is known (section 5.1). Here we are concerned with the nature and energetics of waste product formation. We will outline the main controls and briefly consider the use of faeces and urine as energy and nutrient sources for other organisms.

7.1 Faeces

The composition and quantity of faeces depend on the nature of the ingested food and the extent of digestion and absorption. Digestive mechanisms are diverse, sometimes (primitively) intracellular but more frequently extracellular in a specialized alimentary canal or digestive tract. Initial mechanical breakdown of food is common, especially the crushing of cellulose (indigestible by most animals) by, for example, insect mandibles and mammalian molars. This releases nutritive material otherwise unavailable for hydrolysis by enzymes. The simple molecules resulting from digestion are then taken up by the gut wall, usually by active transport (section 4.5). Faecal material arises in four ways and a scheme illustrating the factors which may control faecal quality and quantity is shown in

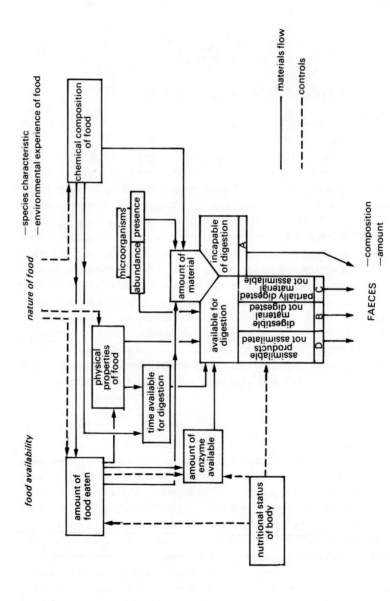

Figure 7.1 The control of faecal production.

figure 7.1. First, some ingested food may not be digested, perhaps because the appropriate enzymes are not present. Secondly, food may not be digested, or be only partially digested, because insufficient enzyme was present or the retention time of food in the alimentary canal was too short. Thirdly, some products of digestion may be incapable of assimilation, their molecular size being too large, for instance. Finally, molecules capable of assimilation may not actually be assimilated. Hence the amount of faeces produced and its composition is determined by a variety of factors, some intrinsic to the animal and some of environmental origin.

Although the majority of species are herbivores, many do not possess the glucosidases necessary for splitting the cellulose which is a major constituent of green plants. Some herbivores have a "microbial flora" present in the alimentary canal and these symbionts break down complex plant polysaccharides. Up to 20% of food intake may be digested by symbiotic microorganisms but lignin is resistant even to this method and its presence also lowers the digestibility of remaining complex carbohydrates. Ruminant mammals are most effective at utilizing green plant material but digest only 50% of the dry weight of their food. As a consequence of the presence of large amounts of dietary cellulose, and commonly a lack of suitable enzymes to digest them, herbivores produce faeces which are a high proportion of their food intake, composed mainly of cellulose and lignin. Whereas the diet of herbivores is primarily carbohydrate, that of carnivores is largely protein and they possess a wide variety of protein-digesting enzymes. Much of the food intake of carnivores is digested and absorbed and their faecal production is low. Some animals feed on highly specific foods. Insects feeding only on the phloem sap of plants, for example aphids, have a diet rich in sucrose but poor in amino acids. Their faeces are mainly sugars ingested in excess of the animal's energy requirements in order to obtain adequate nitrogen. Carnivores which eat their prey whole usually egest a considerable amount of skeletal material. Blood-sucking insects, ticks and leeches produce very few faeces because their food is almost completely digestible and so almost all is assimilated. Herbivores which are specialized seed-eaters may eliminate only 20% of the energy content of their food whereas sap-sucking insects eject 90%. An average figure for herbivores eating whole plants is 65% and for carnivores 20% (i.e. assimilation efficiencies of 35% and 80% respectively; see table 8.2).

Considerable quantities of energy are often egested, therefore, and many organisms exploit this as a food source. Some animals, rabbits for example, eat their own faeces and the larvae of some fleas eat the faeces of their parents. The majority of faecal feeders are insects, some feeding on the microorganisms feeding on the faeces. Many Coleoptera and Diptera feed on faeces, especially those produced by mammals. Ants often collect the faeces of sap-sucking insects and in return they provide protection from predators. Quantitative aspects of faecal feeding will be considered in chapter 8.

7.2 Urine

Animals catabolize (break down) organic molecules during metabolism to release energy and produce simple compounds required for anabolism (construction). Macromolecules for catabolism, derived either from absorbed food or from body tissues, are mainly of carbon, hydrogen, oxygen and nitrogen. Nitrogen-containing molecules are especially important for animals because they are essential for making proteins (e.g. enzymes and muscle proteins) and nucleic acids, necessary for maintenance and growth.

When protein is present in excess of the body's immediate nitrogen requirements, it can be catabolized to release energy, used as a source of carbon for the synthesis of non-nitrogenous molecules, or converted to storage compounds (fats and carbohydrates) to be used eventually as energy sources. The nitrogen is not required for any of these purposes and is removed. Urine production and excretion is concerned with the elimination of nitrogen. The nitrogen in food and muscle is mainly in proteins, which are broken down into amino acids. The amino group ($-NH_2$) of an amino acid is removed by deamination, producing ammonia (NH_3). An example is the deamination of the amino acid glutamate to form α-ketoglutarate:

$$NAD^+ \quad NADH + H^+$$

$$COOH \cdot CH(NH_2) \cdot (CH_2)_2 \cdot COOH + H_2O \rightarrow COOH \cdot CO \cdot (CH_2)_2 \cdot COOH + NH_3 \quad (7.1)$$

The enzyme involved here is glutamate dehydrogenase, NAD^+ (see section 4.2) or $NADP^+$ acting as coenzyme.

When ammonia cannot be eliminated from the body rapidly, it is converted to a less toxic compound. Protein catabolism accounts for over 90% of excreted nitrogen and the remainder results from nucleic acid catabolism. The nitrogen in nucleic acids is contained in purines and pyrimidines, both of which are ring compounds. Pyrimidine nitrogen is a small component of total nitrogen excretion and pyrimidines are usually

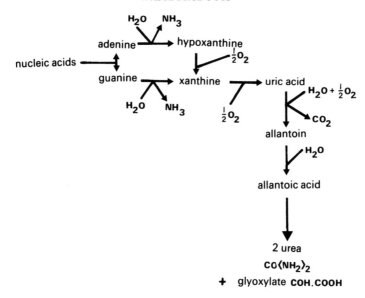

Figure 7.2 The catabolism of purines.

excreted as such. Purine nitrogen is contained in adenine and guanine, double-ring compounds each containing five nitrogen atoms. Purine catabolism is outlined in figure 7.2. Purine nitrogen is eliminated as uric acid by primates, birds, reptiles, some insects and earthworms; as allantoin by most mammals, gastropod molluscs and some insects; as allantoic acid by a few insects; and as urea or ammonia by most fishes, tadpole stages of amphibians, bivalve molluscs, crustaceans and polychaetes.

Excreted nitrogen, then, has two principal sources, being produced in nucleic acid catabolism (where it is eliminated in a variety of forms) and in protein catabolism (where the first and sometimes only form is ammonia). Many animals do not have a supply of water sufficient to allow the dilution and subsequent excretion of ammonia directly, and so these forms convert it to urea or uric acid before elimination. Synthesis of these less toxic compounds requires the expenditure of energy. Ammonia contains nearly 17% of the energy content of the original protein (equation 5.1) and has an energy equivalent of $20.5\,kJ\,g^{-1}$. Although it is therefore a potential energy source for other organisms, ammonium ions released in solution will rapidly form inorganic salts making exploitation by other heterotrophs unlikely.

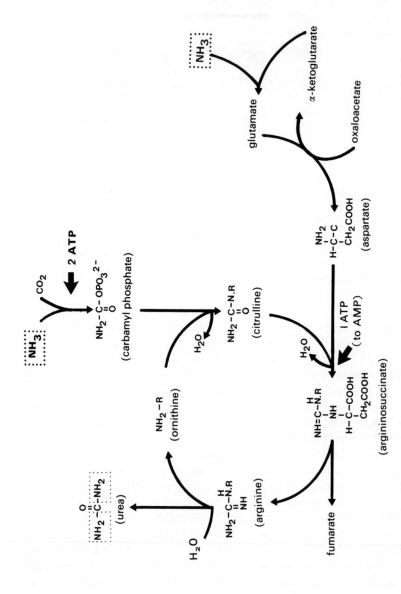

Figure 7.3   The ornithine cycle.

Conversion of ammonia to the less toxic urea is carried out by vertebrates in the ornithine cycle. Urea contains two amino groups, one derived from aspartate which is formed from glutamate. The overall reaction is

$$2NH_3 + CO_2 \rightarrow CO(NH_2)_2 + H_2O \qquad (7.2)$$

and details of the ornithine cycle are shown in figure 7.3. Three ATP are involved in each turn of the cycle. The ornithine cycle may once have served another function, possibly the synthesis of arginine, but animals with inadequate water resources have adapted it for urea production. One mole of urea (about 634 kJ) is produced from 2 moles of ammonia (about 348 kJ each) so the energy value of urea is less than that of the ammonia involved in its formation. If urea is the end product of protein catabolism, a little over 15 % of the original energy content of the protein is eliminated (equation 5.2). Few animals have the enzyme capable of splitting urea (urease) and so it is unlikely to be an important energy source. Ruminant cattle secrete urea, in large amounts, into their alimentary canal, either by direct diffusion from blood to rumen or in the saliva. Urea recycled in this way is an energy and nutrient source for symbiotic bacteria. Urea is often

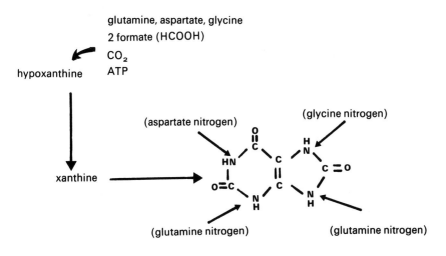

Figure 7.4 Synthesis of uric acid.

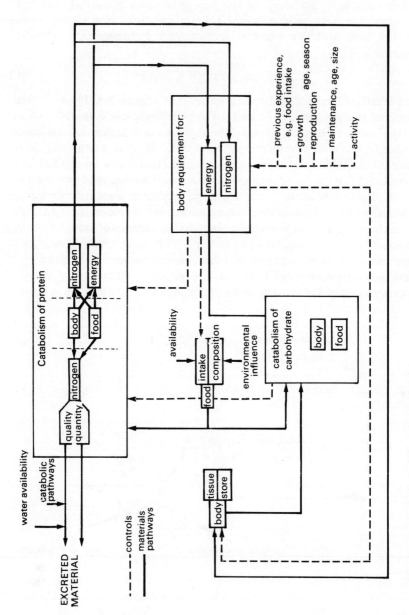

Figure 7.5   The control of nitrogenous excretion.

used as a plant fertilizer, the soil bacteria splitting the molecule and releasing the nitrogen for plant growth.

Uric acid is almost insoluble in water and is produced in large quantities by most animals which experience severe water shortage (for example insects, terrestrial reptiles, and birds). Although uric acid is synthesized during purine degradation, such large quantities are produced by insects and birds that a distinct biosynthetic pathway must be involved. Eight high energy phosphate bonds are required to synthesize purine rings from 5-phospho-D-ribosyl-1-pyrophosphate, which are subsequently converted to uric acid using the metabolic pathways of the early part of purine degradation. The process is outlined in figure 7.4. In birds the xanthine is converted to uric acid by xanthine oxidase but in insects xanthine dehydrogenase is used. Its action is coupled with the reduction of NAD^+ to $NADH + H^+$ which is then oxidized, yielding 6 moles of ATP per mole of hypoxanthine oxidized to uric acid. Whereas uric acid production in birds is expensive in energy terms it is less so in insects because much of the energy expended in purine biosynthesis is recovered. If uric acid is the end product of protein catabolism about 23 % of the original energy content of the protein is incorporated into it (equation 5.3). Uric acid has an energy value of $11.5 \, kJ \, g^{-1}$ but there are few indications that organisms other than bacteria utilize it as an energy source. The large accumulations of guano beneath some bird roosting areas are used by man as a plant fertilizer. Only 6.5 mg of uric acid will dissolve in $100 \, cm^3$ water at 37°C and this extreme insolubility allows it to be stored within the body. The use of uric acid salts for colouring butterfly wings is well known, and some insects store uric acid salts and use them under dietary stress, especially when protein is in short supply. Intracellular gut symbionts are probably responsible for their mobilization and use.

The occurrence of nitrogenous excretory products in animals depends primarily on the availability of water. When water is abundant, ammonia is usually excreted, and when it is not, urea or uric acid are synthesized. There is a metabolic energy cost to the organism in elaborating nitrogenous waste products and when uric acid is produced, a large proportion of the energy in the original protein is eliminated. The utilization of these waste products by other animals is not extensive but a consideration of the magnitude and pattern of nitrogenous excretion is of significance in studying the energetics of individual organisms. Figure 7.5 attempts to outline the more important pathways and controls—only partial quantification has been attempted and only for a few animals of economic importance.

SUMMARY

1. Faeces and urine have separate origins and distinctive chemical compositions. Where possible they should be measured as separate energy pathways, F and U.

2. The composition and quantity of faeces depends on the nature of the ingested food and the extent of digestion and absorption. Considerable quantities of energy are often egested and many organisms exploit this as a food source.

3. Urine production and excretion are concerned with the elimination of nitrogen from the animal's body. Protein catabolism accounts for over 90% of excreted nitrogen, the remainder resulting from nucleic acid catabolism. Amino acid deamination yields ammonia but many animals convert this to urea or uric acid before elimination.

4. Ammonia contains nearly 17% of the energy content of the protein from which it was derived whereas if urea is the end product of catabolism, a little over 15% of the protein energy content is eliminated. If uric acid is formed 23% of the protein energy content is lost. There is also a metabolic energy cost to organisms in elaborating nitrogenous waste products so that they may be voided without toxic effects.

CHAPTER EIGHT

ENERGETICS OF INDIVIDUALS
AND POPULATIONS

In the previous chapters we have considered in turn each component in an energy budget. Such a budget shows how the energy consumed as food (C) is dissipated as heat (R), retained as growth (P, production), or lost in nitrogenous waste (U) and faeces (F). We have also seen how the flow of energy into P or R, for instance, is influenced by a variety of factors. In this chapter and the next we will consider relationships between the budget components, and look at some budgets as a whole.

Studies of the energetics of individual animals and of populations yield information on energy flow within a trophic level. They are considered in this chapter. Information on energy flow through and between two or more trophic levels throws light on the operation of an ecosystem as a whole, and is discussed in chapter 9. These two aspects are distinguished and interrelated in figure 8.1.

8.1 Energy budgets for individual animals

Information on energy budgets forms a bewildering array, but a simple division is between laboratory data, from which a budget can be constructed for individuals living under known (and controlled) conditions, and field data, from which an energy budget for a whole population can be built. Although most studies are performed on a single species, this is not always the case and some energy budgets, derived from community studies, concern broader taxonomic groups (such as an avifauna, or oribatid mites). Almost every energy budget has been constructed from data collected in a unique way, and based on a unique set of assumptions.

Energy budgets are used by ecologists either to evaluate the role of particular animals within a specific habitat or community or to make generalizations which may enhance our broader understanding of eco-

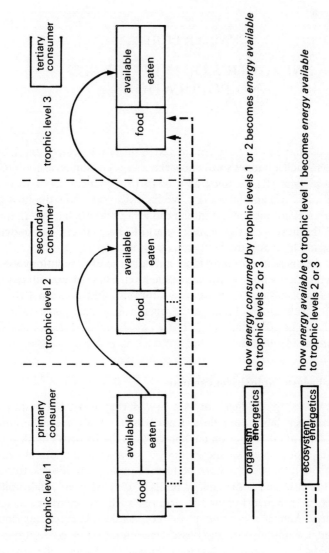

Figure 8.1   The relationship between the energetics of organisms and ecosystems.

system function. Here we shall consider first the value of energy budgets in a broad ecological context and then (section 8.6) view their role in understanding the dynamics of a particular community.

Even if all energy budgets had been constructed using the same assumptions and in the same way, hence reducing or eliminating methodological differences, it would still be difficult to establish useful rules and guidelines. This is because population or laboratory energy budgets will always reflect the idiosyncrasies peculiar to individuals of the species being studied, such as their particular population size, age structure, feeding habits and body size. It almost always makes more sense to compare the ratios of two energy budget components rather than to use raw data. In addition, there is no way to distinguish "reliable" from "unreliable" energy budgets. For these reasons some ecologists believe that there is sufficient "noise" in such data to make valid comparisons and generalizations impossible.

Another problem concerns different concepts of assimilated energy. The energy budget has the form $C = P + R + U + F$ (equation 2.2). Assimilated energy then equals $P + R + U$, for it is the energy absorbed across the gut wall (Calow, 1977). Only a few ecologists (e.g. Ricklefs, 1980) define it in this way, however, the majority following Petrusewicz and Macfadyen (1970) and regarding assimilation as the energy channelled into R and P only. This is mistaken, because the energy lost in nitrogenous waste (U) is as involved in maintaining homeostasis of the organism as is respiration. Ecologists tend to exclude U from A either because it can be difficult to separate nitrogenous waste from faeces, or because of a long-held belief that the amount of energy channelled into U is insignificant and so can be safely ignored. In this chapter, however, we will follow the majority of ecologists and consider A as $P + R$ only, because most ecological data have appeared in this form and it would be impossible to convert them into the more realistic form.

8.2 Energetic efficiencies

It is possible to calculate up to ten different energy ratios using individual energy budget components, and several more if some pathways are combined (e.g. assimilation). Not all the possible ratios are ecologically interesting or meaningful, however. Table 8.1 shows those in general use.

Ecological energy-flow ratios are usually called "efficiencies" and are expressed as percentages. Kozlovsky (1968) and Collier et al. (1973) have catalogued ecological energy efficiencies and the terminology associated

Table 8.1   Useful ecological efficiencies and their common names

P/C—energy coefficient of growth of the first order
     —ecological growth efficiency
     —efficiency of assimilation
     —gross efficiency of growth
     —gross production efficiency

P/A—energy coefficient of growth of the second order
     —growth efficiency
     —tissue growth efficiency
     —net growth efficiency
     —net production efficiency

A/C—assimilation efficiency
     —trophic level energy intake efficiency
     —efficiency of digestion

with these ratios is certainly confusing. Table 8.1 lists the names commonly given to the three most useful ratios from an ecological viewpoint. There is no merit in using any one of these names rather than others, as none is in universal use. Here we will call each measure of ecological efficiency by its appropriate ratio (P/C, A/C or P/A). Some authors have calculated ecological energy efficiencies other than these and the ratios R/C, R/A and F/C are used in some ecology textbooks. It is obviously important to understand the ecological significance of the ratios P/C, A/C and P/A before considering the efficiency of different animal groups, because we must bear in mind that ecologists are interested in the functioning of ecosystems. Therefore we should consider how knowledge of ecological energetics expressed as energy ratios for individuals or for populations (i.e. in the same trophic level) can help in this.

8.3   Production and consumption

P/C indicates how much of the energy consumed by animals is available to the next trophic level. It is clearly of importance to this next trophic level (and to those studying it). The other measures of efficiency, P/A and A/C, simply reflect the mechanisms which underlie the size of P/C and are important solely because they help to tell us why P/C has the size it does in a particular case. A/C expresses how much energy is being extracted from the food consumed, reflecting the quality of food for that particular animal and the adequacy with which its digestive physiology is extracting energy. P/A represents the amount of extracted food energy which is converted to biomass. The ratio R/A is often quoted. It indicates how effective the animal is at converting assimilated energy into a form which can do work, eventually manifest as heat. It must make more sense, ecologically, to consider the amount of assimilated energy passed to the next trophic level (P/A) rather than that passed to the atmosphere (R/A). (The size of P/A will certainly be reduced if R/A increases, but what is ecologically important is the product rather than the energy cost of making it.)

Because these ecological efficiencies are ratios expressed as percentages they can be calculated from various data concerned with feeding and production without these necessarily being expressed as energy equivalents. An estimate of A can be obtained by subtracting F (and possibly U) from C but an estimate of R can only be included if all channels are expressed in energy units. It is not strictly valid to compare biomass budgets with energy budgets, because the energy values of C, F and P are unlikely to be the same. Errors will occur, their magnitude depending upon the energy value differences and the relative size of each energy channel. We shall therefore ignore biomass budgets and their corresponding efficiencies unless they are based upon energy budgets.

Most ecologists compare energy efficiencies by grouping them into categories reflecting feeding method, physiology, habitat and taxonomy. Some distinguish different life styles, for instance separating social from non-social insects. We shall draw a distinction between energy budgets for individual animals and for field populations. It is of interest to see how the age structure of a population, climate and the rigours of predation and competition affect the basic pattern of energetic efficiency. Many workers have not calculated any or all of the ratios P/C, P/A and A/C, and uncommon animals, or those whose habits make experimentation and recording difficult, are poorly represented. The most extensive survey of energy efficiencies in ecological text books is that by Ricklefs (1980), who lists 32 species grouped into heterotherms and homeotherms in aquatic and terrestrial habitats. We will adopt his broad divisions but subdivide aquatic and terrestrial groups into feeding categories. (It is sometimes possible to distinguish herbivores, carnivores, granivores, lactivores, detritivores and parasites.) Humphreys (1979) has collated field reports and analysed 235 energy budgets for natural populations of animals. He calculated only P/A and grouped his data according to its value (placing together those ratios which are similar). Consequently his groupings do not always fit the physiological, habitat and feeding groups previously outlined. Schroeder (1980), who was concerned principally with individuals under laboratory conditions, has analysed 102 literature reports (mostly for species) and has calculated P/C, P/A and A/C for every case. Kucera (1978) and Collier *et al.* (1973) have also made a contribution.

All these have been combined in this chapter to yield a comprehensive yet generalized picture of animal energetic efficiency. It soon became clear that there were considerable and consistent differences between laboratory energy budgets and population energy budgets. Twenty-two efficiency ratios have been calculated, for both individuals and populations, and in 16 there were differences greater than 10%. It is unfortunate that these individual and population data cannot be combined, as together they would form an impressive body of knowledge. On the other hand, the very fact of consistent differences between laboratory individuals and field populations indicates that parameters other than those moulding individual animal efficiencies are significant in determining population energetic efficiency. This is an important point, to which we shall return later.

8.4 Energy efficiency of an individual animal

Energetic efficiencies according to feeding type and habitat are given in table 8.2. The numbers of species used to calculate the means are variable; more data are available for herbivores than for carnivores, and data for homeotherms are sparse. The ratio A/C indicates the proportion of food energy which is absorbed by the gut and so becomes available for

Table 8.2 Individual animal energy efficiencies (after Schroeder, 1980). A—assimilation, P—production, C—consumption. Figures in brackets show numbers of species involved.

			A/C	P/A	P/C
poikilotherms **(heterotherms)**					
herbivores	terrestrial	(32)	45%	52%	20%
	aquatic	(15)	61%	56%	34%
granivores	terrestrial	(4)	78%	30%	24%
carnivores	terrestrial	(11)	84%	58%	46%
	aquatic	(17)	64%	48%	30%
detritivores	terrestrial	(6)	12%	50%	6%
	aquatic	(6)	44%	56%	25%
parasites		(3)	77%	50%	42%
homeotherms					
herbivores	terrestrial	(3)	66%	23%	13%
granivores	terrestrial	(3)	76%	29%	22%
lactivores	terrestrial	(2)	95%	45%	43%
carnivores	terrestrial	(—)	—	—	—

production. The most important factor determining the size of A/C is the nature of the food. A/C is relatively low in herbivores (57%) because plants contain a large amount of cellulose which most animals cannot digest, and the energy in the cellulose is therefore not available to them. Carnivores have a diet higher in protein and lower in carbohydrate and their A/C is, on average, 80%. Insectivores, which are a type of carnivore, have an A/C ratio lower than expected because their food contains a significant amount of chitin, which is almost totally indigestible. If the exoskeleton is not eaten, however, then A/C should be higher. Homeotherms with the same food type as heterotherms sometimes have a higher A/C. We do not know why this is but it could be due to specialized features, such as gut fermentation of cellulose for instance, which may be more easily controlled in the stable internal environment of a homeotherm. Even within these broad feeding categories there is great variation in A/C, reflecting diverse feeding strategies. A herbivore, for instance, may eat only leaves or stems or seeds. It may extract the contents of phloem, xylem or parenchyma cells. Each of these is different in energy value and chemical composition. The ratio A/C is much more variable than P/A and has an important influence on P/C.

Once energy has been assimilated, the magnitude of P/C depends entirely on P/A, which reflects how much energy is used for body maintenance and activity and how much is laid down as biomass or production. Several authors have calculated the theoretically best possible P/A efficiency. Calow (1977) suggests that a growing heterotroph system could achieve a P/A ratio of between 70% and 80%, and Schroeder (1980) quotes a maximum of 88%. Most people believe that the P/A ratio should be lower in carnivores when compared with herbivores. Carnivores have to expend energy searching for and capturing food and as a consequence their respiration expenditure is likely to be higher than a more sedentary herbivore of the same size. P/A should also be lower for homeotherms when compared with heterotherms. The maintenance by homeotherms of a constant body temperature, above the ambient, requires an increase in respiratory energy costs. Many animals' life styles are energy-intensive, locomotion (and other activities such as nest-building) requiring considerable respiratory energy expenditure (section 5.2). Such policies will also lower P/A, as will environmental stresses such as temperature extremes, shortage of food or nutrient imbalance. Under these conditions the animal either has to behave in a way which may cost energy, or production becomes limited by factors other than energy availability (such as a requirement for specific amino acids). Table 8.2 indicates that P/A is rather unresponsive to food and habitat differences, there being no trend discernible. This may be a consequence of laboratory energy budgets, where the animal is rarely made to hunt for food, for example. There is a clear distinction between poikilotherms (heterotherms) and homeotherms, however, even under laboratory conditions. The mean P/A for poikilotherms is 50% whereas for homeotherms it is only 32%, presumably because the latter have a relatively higher metabolic rate (a higher value for R), which reduces the proportion of A which can go into P.

The few observations for many of the feeding categories make broad generalizations very speculative. Although the overall mean P/C for aquatic animals is 30% as opposed to 22% for terrestrial forms, this should not be taken as an indication that life on land is energetically more costly. The difference disappears if the low P/C for terrestrial detritivores (a consequence of their food quality in all probability) is excluded. Indeed P/A ratios for aquatic and terrestrial animals are broadly similar, indicating that the maintenance costs of these groups are not greatly different. An animal feeding on plant material passes on less of its energy intake to the primary carnivore trophic level than the latter does to the secondary or tertiary carnivore levels. The mean figures of 21% for herbivores and 41% for carnivores indicate the scale of the difference but because of the great variability of the data this cannot qualify as a valid generalization. The most "desirable" energy flow pattern through ecosystems (from the viewpoint of mankind) will depend upon those ecosystem characteristics which he wishes to preserve or enhance and this will be discussed in chapters 9 and 10. A particular flow of energy through individual animals can of course be achieved through changes in either A/C or P/A, or both, but the energy flow of greatest benefit to other consumer organisms is that with the lowest P/A. Only energy lost as heat during respiration is without further ecological significance.

8.5 Energy efficiency of a population

It is of considerable interest to see how the pattern of individual animal energy flow compares with that of a population, where the influences of a varied age structure, reproductive status, the physical environment and variable food quality and quantity have been experienced. Energy efficiencies given in table 8.3 are lower than the equivalent values for individual animals in table 8.2. The mean A/C for all categories of individual animals in the laboratory is 64%, but their collective field performance as a population is 52%. It seems that the average animal

Table 8.3 Population energy efficiencies. (Symbols as in table 8.2. Figures in brackets show number of species involved.)

		A/C	P/A	P/C
poikilotherms (heterotherms)				
herbivores	terrestrial	32% (5)	37% (69)	14% (4)
	aquatic	47% (4)	26% (4)	11% (5)
granivores	terrestrial	1% (1)	—	—
carnivores	terrestrial	77% (2)	36% (19)	30% (2)
	aquatic	—	—	—
detritivores	terrestrial	35% (3)	13% (1)	6% (1)
	aquatic	45% (1)	32% (3)	6% (1)
parasites		—	—	—
homeotherms				
herbivores	terrestrial	65% (10)	3% (16)	2% (8)
granivores	terrestrial	81% (1)	4% (2)	4% (1)
lactivores	terrestrial	—	—	—
carnivores	terrestrial	88% (4)	2% (8)	2% (4)

living in the field extracts 12% less energy from its food than does its laboratory counterpart. The immediate conclusion is that food in the field is not markedly different in its digestibility from that provided in the laboratory. When we consider P/A, however, a different picture emerges. Field populations turn only 19% of their assimilated energy into production whereas those in the laboratory achieve 45%. This is a stark indication of the energy cost of living in the "real world". The average animal population in its natural ecosystem generates an energy flow of only 9% for the use of the next trophic level (P/C), compared to the average of 28% for individuals living in the laboratory.

Some authors have tried to enhance the ecological meaning of energy ratios for populations by plotting one series of ratios against another. Welch (1968), Schroeder (1980), McNaughton and Wolf (1973), Ricklefs (1980) and Kucera (1978) have all plotted A/C (indicating the quality of food consumed and the effectiveness with which energy is extracted by digestion and absorption) against P/A (indicating potential for growth and reproduction) for a variety of populations. Figure 8.2*a* shows that Welch (1968) was able to present a clear picture for aquatic animals, with P/A negatively correlated with A/C, and concluded that carnivores tended to have a higher A/C and a lower P/A. Thus herbivores and detritivores assimilated less energy from their food (possibly because it was of poor quality) but used more of what they acquired for growth (perhaps a lower

R because they are more sedentary). The data used by Welch (1968) appear to come principally from laboratory studies, some of which were not based on energy units. By 1980, Schroeder had more extensive data available, and all in energy terms. He shows (figure 8.2*b*) a shallow

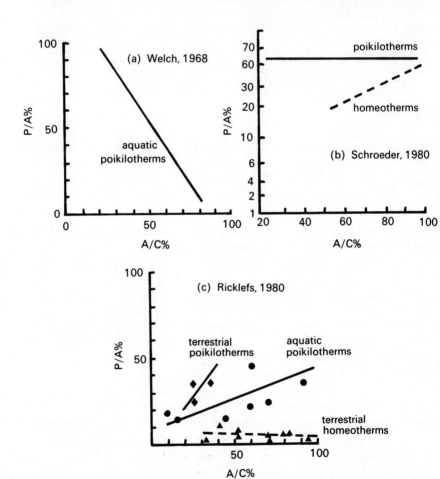

Figure 8.2 The ratio A/C plotted against P/A for various groups of animals, by various authors.

(*a*) Welch (1968). Energy, weight and carbon data. Laboratory animals. 28 observations.

(*b*) Schroeder (1980). Energy data. Laboratory data. 3 homeotherm and 12 poikilotherm observations.

(*c*) Ricklefs (1980). Energy data. Field populations. 7 aquatic and 3 terrestrial poikilotherm and 9 terrestrial homeotherm observations. (Lines fitted by eye.)

negative correlation for poikilotherms (the approximate equivalent of Welch's aquatic consumers) but a steep positive correlation (based on very sparse data) for homeotherms. The conclusion of these two authors is that as food quality or the effectiveness of digestion increase, proportionately less assimilated energy is channelled into production, but the flimsy evidence for homeotherms conflicts with this.

Ricklefs (1980), taking the same approach, and using only population energy data, comes to a different conclusion (figure 8.2c). Populations of aquatic poikilotherms show a positive correlation between P/A and A/C, that is to say as food quality and digestibility increase, maintenance costs are reduced and productivity is higher. Terrestrial poikilotherms, however, have a weak negative correlation with P/A, considerably lower than that found by Schroeder (1980) for laboratory animals. This is consistent with the previous comments concerning the performance of field populations. The reverse relationship between A/C and P/A for field populations and laboratory animals is more difficult to explain. The fact that three variables (P, R and C) are involved in these relationships does not make the task any easier and we must conclude that at the moment arranging data in this way does not really enhance our understanding of population or individual energetics.

Welch (1968) has stressed the *predictive* value of linear relationships between P/A and A/C. If P and R can be measured (and this is relatively easy) then P/A can be calculated and A/C interpolated from any linear relationship between the two. This enables estimations of C and of F + U without their being directly measured (according to Welch a relatively more difficult task). For example, if P = 40 kJ per unit time, say, and R = 80 kJ, then

$$\frac{P}{A} = \frac{40}{80+40} = 33\%$$

If the relationship between P/A and A/C was found to be such that a value of 33 % for P/A gave a value of 40 % for A/C, then

$$\frac{A}{C} = \frac{80+40}{C} = 40\%$$

and therefore C = 300 kJ.

As C = P + R + U + F (equation 2.2), then

$$300 = 40 + 80 + U + F$$

and

$$U + F = 180 \, kJ.$$

Evidence presently available does not indicate a simple linear relationship between P/A and A/C, however, and it is not easy to predict one on theoretical grounds. In these circumstances it is difficult to accept the predictive value of any relationship between P/A and A/C and there is little reason to think that it is of any general ecological significance.

8.6 The construction and ecological value of population energy budgets

The generalizations outlined in the previous section concerning energy flow in animal populations are only as sound as the energy budget data upon which they are based, and this chapter would be incomplete without a consideration of how population energy budgets are constructed. In addition, such budgets are a valuable ecological tool in their own right, because they bring together information and ideas from the diverse fields of population dynamics and physiology. In this section we shall illustrate both points by using one of the classical studies of animal population energetics, that by Wiegert (1964) on meadow spittle bugs (*Philaenus spumarius*) in Michigan, USA. This insect belongs to the order Homoptera and feeds by sucking xylem sap of a wide variety of herbaceous plants including many leguminous crops. It produces a single generation annually and overwinters as eggs. The larvae secrete a bubble-filled fluid (spittle) and moulting between each of the five instars takes place within the spittle mass. The adults live for several months, females eventually ovipositing during September and dying in October. Wiegert studied populations living in "Old Field" (an abandoned farming area) and also in adjacent hayfields of alfalfa (*Medicago sativa*) which were cut twice each summer. When the alfalfa was cut there was a mass exodus of adults to the Old Field, many of whom recolonized the crop before its second harvest. We will concentrate only on those populations inhabiting the Old Field where there was both a natural population and an influx of adults at haymaking.

Population energy budgets are usually constructed by combining laboratory data on individual animal growth, egestion, respiration and consumption, acquired over a range of temperatures, with field data on the number and biomass of each life stage. The precise method of construction varies with the particular characteristics of the life cycle but a general scheme is shown in figure 8.3. The task for Wiegert was relatively simple as spittle bugs have only a single generation each year, but their methods of feeding and egestion made it difficult to quantify these energy channels. The first stage in constructing a population energy budget leads to the formation of *individual* animal energy budgets for a particular period, a

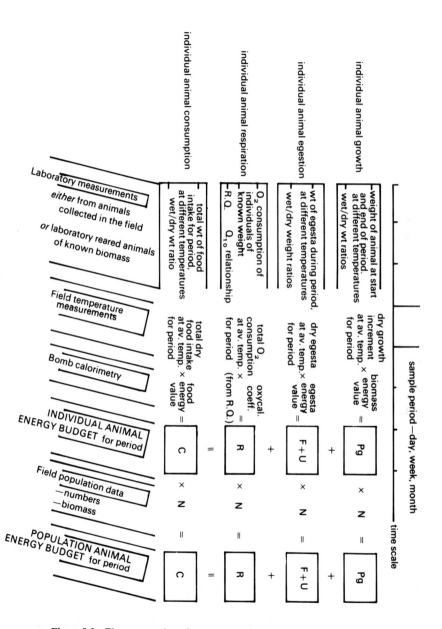

Figure 8.3 The construction of an energy budget.

Table 8.4   Energy budgets for individual *Philaenus spumarius* (joules day^{-1})
Respiration, growth and egestion are independent estimates and are summed to yield second estimate of consumption (modified from Wiegert, 1964)

4th instar nymph (1.16 mg wet weight) at 25°C			
			A/C
Independent estimate of ingestion (C)		8.79–28.87	
Respiration (R) 3.18			
Growth (Pg) 3.10 } (C)		10.96–20.84	30–57%
Egestion (F + U) (4.69–14.56)			

Adult (3.87 mg wet weight) at 21°C			
			A/C
Independent estimate of ingestion (C)		7.53–24.69	
Respiration (R) 4.18			
Growth (Pg) — } (C)		5.90	71%
Egestion (F + U) 1.72			

day in the case of the spittlebug study. These are shown for two life stages in table 8.4. Most of the fluid ingested by these animals is subsequently egested as spittle by nymphs and expelled from the anus by adults. Such material was collected and its dry weight, energy value and ash content determined. An independent estimate of ingestion was obtained by collecting tomato xylem sap exuding from cut stem stumps, determining its energy value per unit dry weight and multiplying this by the dry weight of spittle or anal fluid collected from animals also kept on tomatoes in a high humidity. (Tomato plants were used because their cut stem stumps exude xylem sap whereas alfalfa plants would not.) *Population* energy budgets are then produced by multiplying data for individual animals by the numbers or biomass of the field population and summing for all sampling periods throughout the season (figure 8.3).

The example of the spittlebug population study given in table 8.5 shows the energy budget for the Old Field in 1959. The particular characteristics of this budget reflect to a large extent the condition of the population during the period studied. Assimilated energy in the Old Field in 1959 was almost 2500 J m^{-2} whereas in 1960 it was over 5000 J m^{-2}, a difference produced by a change in population density. The sort of information yielded by this type of study comes to light when we consider the A/C ratios for nymphs and for adults. The bulk of the energy ingested by spittlebugs is in amino acids present in the xylem sap, and the wide

Table 8.5 Population energy budget for *Philaenus spumarius* in the Old Field, Michigan, 1959 (joules m^{-2} yr^{-1}) (modified from Wiegert, 1964)

Nymphal growth (Pg)	192.5
Nymphal exuvia (Pe)	8.4
Adult egg production (Pr)	16.7
Total production (P)	217.6
Nymphal respiration	159.0
Adult respiration	2075.3
Total respiration (R)	2234.3
Nymphal egestion	723.8
Adult egestion	1037.6
Total egestion (F + U)	1761.4
Nymphal assimilation	359.9
Adult assimilation	2092.0
Total assimilation (A)	2441.9
Nymphal ingestion	1083.7
Adult ingestion	3129.6
Total ingestion (C)	4213.3

	A/C	P/A	P/C
Nymphs	33%	56%	18%
Adults	67%	1%	1%
Whole life	58%	9%	5%

difference between the A/C ratio for nymphs (33%) and that for adults (67%) is therefore surprising. It represents the amount of "surface tension depressant" which nymphs add to their egesta to produce the frothy spittle. Amino acids are usually considered a valuable resource for herbivorous insects and the fact that so much of this food material is expended in spittle production suggests that the spittle must be of considerable survival value to the insect. If Wiegert had used the definition of A/C advocated in section 8.1, then the "surface tension depressants" added to spittle would count as assimilated energy because they are derived from material (and hence energy) which has passed across the gut wall. The A/C ratios for both nymphs and adults would then be much higher and the only non-assimilated energy would be from those amino acids which had not been absorbed. Another aspect emerges if we compare the energy utilization of Wiegert's spittle bugs with that of aphids. Aphids are also sap-sucking feeders but utilize phloem rather than xylem. Their food supply is so rich in sucrose that they egest large amounts of energy-rich material and consequently have very low A/C ratios (Llewellyn, 1972).

SUMMARY

1. The diverse and unique ways in which energy budgets are constructed mean that animal energy data can only be meaningfully compared by calculating energy flow ratios between different channels. These are usually called efficiencies and are expressed as percentages.

2. The three most useful animal energy ratios are P/C, A/C and P/A. P/C indicates how much of the energy consumed by animals becomes available to the next trophic level and thus is ecologically the most important. P/A and A/C reflect the mechanisms which underlie the magnitude of P/C. A/C indicates how much energy is being extracted from the animal's food and reflects both the quality of the food and the effectiveness of the digestive physiology of the animal. P/A represents the amount of extracted food energy which is subsequently converted to animal biomass and its value reflects the metabolic "cost of living".

3. When animals are reared in the laboratory, A/C is relatively low for herbivores (an average of 57%) because plants contain large amounts of cellulose which most animals cannot digest. Carnivores, with a diet rich in protein, have an average A/C ratio of 80%.

4. The P/A ratio has a theoretical maximum of 88%. In reality the ratio should be lower in carnivores than it is in herbivores, because they have expended energy in searching for and capturing prey and consequently less is available for biomass production. Homeotherms, which maintain a body temperature above the ambient, will also have a lower P/A ratio when compared with poikilotherms. The average P/A value for poikilotherms is 50% and for homeotherms 32%.

5. The overall average P/C ratio for aquatic animals is 30% as opposed to 22% for terrestrial forms. This should not be taken as an indication that life on land is energetically more costly, however, as the difference disappears if the low P/C for detritivores (a consequence of food quality) is excluded.

6. Animals in the field have A/C ratios somewhat lower than their laboratory counterparts, extracting 12% less energy from their food. Field populations turn only 19% of their assimilated energy into production, whereas those in the laboratory average 45%. This is another indication of the energy cost of living in the "real world". The average animal population living in its natural ecosystem generates an energy flow of only 9% for the use of the next trophic level (P/C), compared to an average of 28% for individuals living in the laboratory.

7. Population energy budgets are usually constructed by combining laboratory data on individual animal growth, egestion, respiration and consumption acquired over a range of temperatures. The precise method varies with particular characteristics of the animal's life cycle: the study of meadow spittle bugs was difficult because they are fluid feeders, and an independent estimate of C could only be obtained from animals living on an alternative and unusual food supply. The major differences in annual population energy flow of spittle bugs were produced almost exclusively by changes in population density.

CHAPTER NINE

ECOSYSTEM ENERGETICS

The concept of the food chain is a familiar one. It is part of our everyday experience and it is a simple step to the realization that organisms are either primary producers or consumers. The flow of energy, packaged as food items, in biological communities creates a structure influenced by a wide variety of biological and physical factors and itself influencing other biological and physical phenomena. The idea of a food chain with organisms neatly arranged in trophic levels (plant, herbivore, carnivore) is an oversimplification, however. Although herbivores are often specialist feeders taking energy from only one source, this is not true of all trophic levels and many carnivores not only feed widely on other animals but will also take plant food. Consequently trophic relations within communities and ecosystems are a matrix or web, with organisms not restricted to one type of food or to one trophic level. Energy contained in faecal and nitrogenous excretory material is a source of food for some organisms (chapter 7) and most plants and many animals die from causes other than predation by the next trophic level. The concept of the food chain takes little account of this and we need a more comprehensive and realistic framework in which to consider ecosystem energetics.

9.1 Ecosystem energy models

In 1927 Charles Elton suggested that each trophic level could be represented by a rectangle whose size corresponded to the number of organisms at that trophic level. If these blocks are stacked upon each other with the primary producers at the bottom and the final carnivores at the top, then a pyramid results (figure 9.1a). The proportions of the pyramid vary from community to community or ecosystem to ecosystem depending upon the specialized characteristics of the environment. Similar pyramids can

116

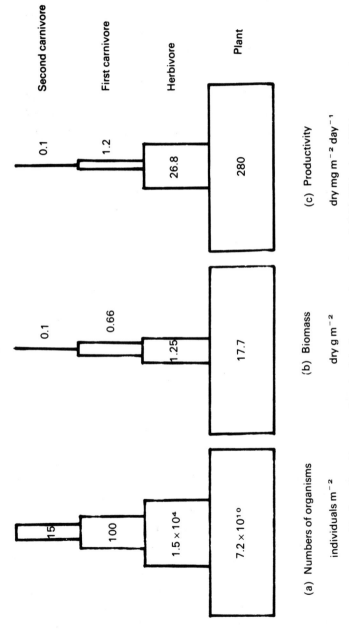

Figure 9.1 Pyramids of numbers, biomass and productivity for an experimental pond. (After Whittaker, 1975.)

be made using biomass (figure 9.1b) or some estimate of productivity (figure 9.1c). They show a decline at higher trophic level which is an inevitable consequence of P/C ratios which are less than 100% (section 8.4) and of the obvious fact that not all plants and animals are eaten by the trophic level immediately above them.

Eltonian pyramids are not adequate models of ecosystem trophic relationships, however, because they do not depict the presence and activity of those organisms that feed on dead plant and animal remains. We shall call these organisms detritivores (feeding on detritus or fragments of decomposing organisms). They are mostly invertebrate animals (earthworms, Collembola and mites) and bacteria and fungi, and occupy a trophic position similar to that of the herbivores in that they support a wide assemblage of carnivores. Detritivores are responsible for the process of decomposition, whereby complex organic molecules built up by organisms are broken down to release carbon dioxide, water and the inorganic salts necessary for plant growth. In some ecosystems, such as woodland, more than 90% of the primary production may be consumed by detritivores and less than 10% by herbivores.

Many ecologists exclude bacteria and fungi from the detritivore trophic level and consider that only animals feeding on dead remains are detritivores. It is suggested that these "animal detritivores" play an important role in decomposition—their activity speeds the decay of litter by microorganisms because they macerate dead leaves in their digestive tracts. This exposes a greater surface area of dead plant material upon which microorganisms can feed. Ricklefs (1980) argues that although detritivores do perform this function they are no different from other animals who ultimately return most ingested food to the environment as water, carbon dioxide, and mineral salts excreted in sweat and urine—they are not readily distinguishable from other organisms in the way they use their energy intake. While Ricklefs' points are valid it can be argued that detritivores *do* have a unique role in enhancing cycling of inorganic plant nutrients in ecosystems, not because they are physiologically special but because of the magnitude of their activity. The relationship between "animal detritivores" and microorganism detritivores is complex, however. Both groups obtain their energy and nutrients from the same source and therefore should be classed in the same trophic level. But when "animal detritivores" feed they must inevitably consume some bacteria and fungi already feeding on the dead material, and may obtain some energy from digesting them. Clearly the trophic level approach to ecosystem energetics is equivocal in this instance but we do not yet have a clear enough

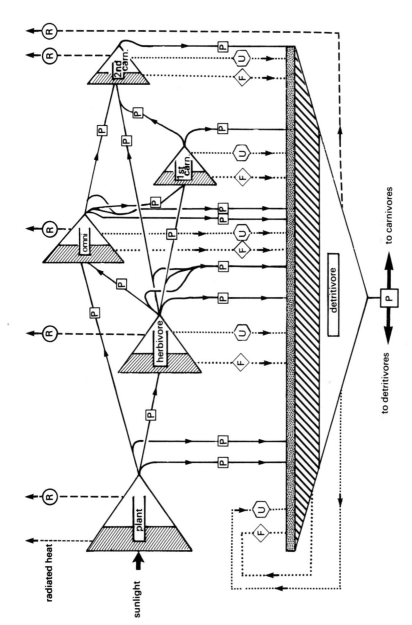

Figure 9.2 A simplified model of ecosystem energy flow.

understanding of the energetics of decomposition to suggest an alternative scheme.

A more comprehensive model of energy flow through ecosystems is shown in figure 9.2. The scheme includes autotrophs (plants), primary consumers (herbivores), secondary and tertiary consumers (primary and secondary carnivores) and omnivores (animals feeding at two trophic levels, one of which is autotrophs). Detritivores are shown as one group, irrespective of their taxonomic affinity or precise role in decomposition. The A/C ratio for each trophic level is represented by a line vertically bisecting each triangle and in plants this is referred to as *gross primary production*. F and U from all trophic levels immediately become available to the detritivores, but P is divided between that captured by predators (energy moving to the right across the model) and that which is available to detritivores. Material available to detritivores may not be used immediately but may accumulate in or on the soil or in aquatic deposits. This is indicated in figure 9.2 as a rectangle adjacent to the input side of the detritivore triangle, and will be called the *detritus store*.

Energy from sunlight captured by plants is eventually either lost as heat to the atmosphere, as a result of metabolic processes, or accumulated in the organic stores—those of herbivores, carnivores and omnivores in the short term, and the detritus store in the longer term. The principal deficiency of this model is the same as one of those of Elton's trophic level concept: populations and organisms do not always fit easily and sensibly into a particular trophic level. (Is a parasite simply a specialized carnivore? Where should we place omnivores?) This model has a triangle for omnivores but we could equally well assign part of their energy flow to the herbivore level and part to the carnivore level. It also illustrates why ecologists have shown such interest in the production pathway (P). There are 22 pathways which production can follow and this makes it a very versatile "energy package" which can produce considerable ecosystem diversity. F and U, on the other hand, have only ten pathways between them because they travel only to the detritivore triangle. Perhaps this picture will change when the detritivore trophic level is better understood, and not merely used by ecologists as a dumping ground for non-living "energy packages".

9.2   Ecosystem efficiency

Because all living organisms require energy, it is possible that the numbers of animals in natural populations and ecosystems, and their ways of life,

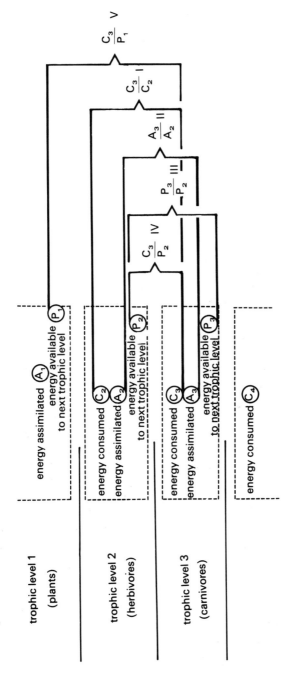

Figure 9.3 Ecological efficiencies for ecosystem studies.

are determined by the way in which they handle energy. We should be able to predict the size and number of trophic levels present in any community and ecosystem by considering the proportional energy flow between different elements, that is by studying energy efficiencies. The movement of energy through an animal depends on the efficiency with which it assimilates energy from its food (A/C) and the efficiency with which this assimilated energy is turned into production (P/A) (chapter 8). The movement of energy through an ecosystem, however, depends on the efficiency with which populations exploit their food and then convert it into biomass. The ratio of production to consumption for individual animals forms the basis of the efficiency of energy flow in ecosystems. We now consider wider ecological relationships between the organism and its environment.

Figure 9.3 shows an ecosystem with three trophic levels (plants, herbivores and carnivores) and indicates the various energy flow ratios or efficiencies that can be calculated. Energy flow in individual organisms occurs inside the boxes but in the context of the whole ecosystem, the energy of production (P_1, P_2 and P_3) is available to the next trophic level. Three efficiencies can be calculated to indicate energy flow from one trophic level to the next (I, II and III). A further efficiency value deals with energy flow between all three trophic levels (V). Ecosystem efficiencies were first used by Raymond Lindeman in 1942, in his study of Cedar Bog Lake, Minnesota, USA, where he calculated the ratio A_3/A_2 (II in figure 9.3), which has become known to many as the *Lindeman efficiency* (Colinvaux, 1972). At first sight the ratios P_3/P_2 (III) and C_3/C_2 (I) appear to be similar to A_3/A_2, simply representing the same ratio calculated one step later in the energy flow, but this is not the case. In figure 9.4 we see the components of each of these ratios and they are not the same. Each efficiency depends upon a unique energy flow and therefore each must represent a different facet of energy flow between the two trophic levels. The ratio C_3/P_2 (IV in figure 9.3), the exploitation efficiency of trophic level 3 with respect to P_2, is common to all three ratios, but C_3/C_2 is unique because it incorporates A_2/C_2. P_3/P_2 is unique because P_3/A_3 is incorporated, and A_3/A_2 is unique because it has C_3/P_2 in common but derives P_2/A_2 from C_3/C_2 and A_3/C_3 from P_3/P_2.

What then does each efficiency indicate? Are they all of the same ecological significance? All are commenting upon the transfer of energy from trophic level 2 to trophic level 3 (figure 9.3) but the ratio P_3/P_2 is particularly important because it spans three trophic levels. It can indicate, for instance, how much of the production of trophic level 2 is at least

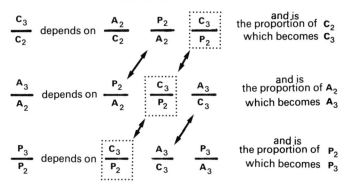

Figure 9.4 The relationship between the ratios C_3/C_2, A_3/A_2 and P_3/P_2. (See text for explanation.)

available as food for trophic level 4. Each of the C_3/C_2, A_3/A_2 and P_3/P_2 ratios is made up of four energy compartments combined as three ratios (figure 9.4) and this makes it exceedingly difficult to interpret their biological significance. Although Kozlovsky (1968) produced an extensive catalogue of ecosystem efficiencies, he failed to establish clearly their ecological significance. There is a widely-held belief (see Colinvaux, 1972) that the ratio A_3/A_2 is preferable because the gross productivity (assimilation) of primary producers can be incorporated into it, but an equally convincing case can be made for P_3/P_2. In practical field studies consumption is very difficult to measure accurately and this usually precludes calculation of C_3/C_2. Many authors do not define precisely which ratio they are considering and some treat them as exactly comparable, which they certainly are not, so any generalizations made as a consequence of comparing ecosystem energy ratios must be treated with caution.

9.3   Lindeman's study of Cedar Bog Lake

Lindeman (1942) studied Cedar Bog Lake for five years and collected data on the standing crop for each trophic level. The turnover rate of the major trophic levels was estimated using the published data of others: phytoplankton populations were replicated (turnover) once each week in summer, zooplankton once every two weeks, rooted pond weeds once each year. Lindeman multiplied standing crop by turnover time to yield production (P) and added an amount to take account of predation and death. Respiration (R) for this total production was calculated by using published data for laboratory animals. A summary of Lindeman's Cedar Bog study is given in table 9.1 together with his ecosystem energy efficiencies. Slobodkin (1962) has pointed out that Lindeman made several mistakes, counting some parts of P and R twice for instance, but in spite of this his conclusions seem reasonable and still provide a basis for generalizations about the trophic structure of communities. The conclusion which Lindeman himself came to, which is most widely quoted, is that consumers at progressively higher trophic levels in the food cycle are more efficient in the use of their food supply. We will discuss the validity of this claim in the light of other (and more complete) ecosystem studies later. Lindeman's study has led to a further generalization which is typified by the statement of Colinvaux (1972) that "typical ecological efficiencies of an animal trophic level might be something like 10 per cent".

Lindeman's method of equating the harvest or standing crop at the end of a specific time period, with production has been used by many workers. It is an attractive proposition because it allows easy calculation of P_3/P_2, whereas in order to calculate A_3/A_2, respiration (R) has also to be measured. Lindeman recognized the problem of accurately estimating losses due to predation and disease, however, and when attempts are made to overcome this the study becomes considerably more complex (as we shall shortly see). There have been many ecosystem studies since Lindeman's pioneering work and we shall now consider one of the more recent.

9.4 The energetics of Cone Spring

The elucidation of energy flow through an ecosystem remains a considerable undertaking and has still been attempted only for simple aquatic habitats. Tilly (1968) investigated Cone Spring, Iowa, USA, a cold freshwater spring of 141 m² and between 0.5 and 16.5 cm deep. Freshwater springs have been popular study areas (Silver Springs, Florida (Odum, 1957); Root Spring, Massachusetts (Teal, 1957), for example) because they have less variable physical factors than terrestrial ecosystems and because their biological diversity is usually relatively small.

Over fifty species of plants and animals were found in Cone Spring, but some were combined into taxonomic groups, usually families. Population data were collected for all species but production and respiration were estimated for only the most abundant. A feature of the Cone Spring ecosystem is that the primary consumers are detritivores rather than herbivores. The trophic levels are therefore detritus (acting as primary

Table 9.1 Lindeman's study of Cedar Bog Lake $(J m^{-2} yr^{-1})$. Calculated from Ricklefs (1980) after Lindeman (1942)

	Trophic level 1 Primary producers	Trophic level 2 Primary consumers	Trophic level 3 Secondary consumers
Production (P)*	3678	435	54
Respiration (R)	979	184	75
P + R (= part A)	4657	619	129
P_2/P_1	11.8%		
P_3/P_2	12.5%		
A_2/A_1	13.3%		
A_3/A_2	20.9%		

* includes that removed by predation

producer), primary consumers (detritivores) and secondary and tertiary consumers (carnivores). There was some import and export as the study area was flowing water. In order that the food web can be simply represented we will consider only ten animal species (figure 9.5), but these were responsible for over 90% of the energy flow. Primary production and detritus import were carefully studied but here the details have been omitted. The Cone Spring study provides a good illustration of the problems and complexities of quantifying food webs in energetic terms. Species are in boxes, arranged in a pattern of feeding hierarchy. For each population, production (P) and respiration (R) were estimated and summed to yield assimilation (A). Consumption (C), faecal production (F) and nitrogenous excretion (U) were not quantified in energetic terms. For each population represented in figure 9.5, assimilation is the upper number and production (energy available for the next trophic level) the lower one. Values for R (which are A minus P) have been omitted to avoid complication. Although figure 9.5 suggests that all consumable production by some species is eaten by carnivores, this is not the case, because energy pathways of production represented by dead consumers have been omitted for clarity. In no instance during the study, however, was the fate of consumer production divided between that consumed by carnivores and that channelled into the detritus store.

Detritus available for consumption by animal detritivores (primary consumers) and microorganism detritivores contained $39.78\ MJ\ m^{-2}$ during the year of the study. $9.98\ MJ$ were estimated to have been consumed by animal detritivores, and $17.15\ MJ$ were exported. Presumably the remainder $(12.65\ MJ)$ was consumed by microorganism detritivores

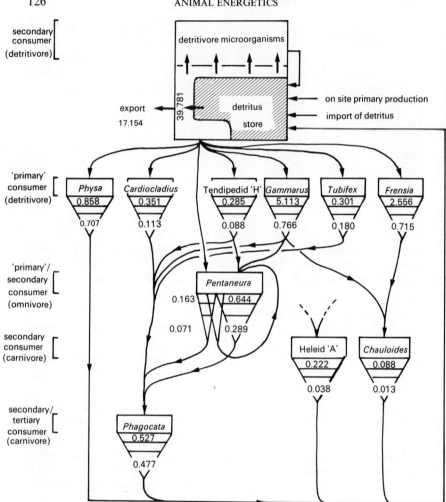

Figure 9.5   Tilly's study of Cone Spring energetics—units $MJ\,m^{-2}\,yr^{-1}$.

(bacteria, fungi and Protozoa) or stored. By far the most important animal detritivore was the crustacean *Gammarus pseudolimneus*. Its assimilation was 54% of the total for purely detritivore feeders but production was only 30% of the total. Its P/A ratio is 15%, indicating that little of the assimilated energy became available to the next trophic level. *Frensia missa*, a trichopteran insect, was the next most important species. Assimilation was 2.556 MJ m^{-2} yr^{-1} and production 0.715 MJ m^{-2} yr^{-2} yielding a P/A ratio of 28%. *Physa integra* is a pulmonate snail and as well as assimilating 0.858 MJ m^{-2} yr^{-1} it plays an important part in making detritus available to smaller detritivores by physically breaking up large food particles. *Cardiocladius* sp. and tendipedid "H" are dipteran insect larvae and *Tubifex tubifex* is an annelid. Between them they assimilate 0.937 MJ m^{-2} yr^{-1} and have a production of 0.38 MJ m^{-2} yr^{-1}. (Tendipedid "H" was one of several species of that genus which were not completely identified.) Apparently *Physa* is not preyed upon by any resident of Cone Spring. *Cardiocladius* is eaten solely by *Phagocata velata*, a planarian worm, and *Frensia* is eaten only by *Chauloides pectinicomis*, a megalopteran insect present as the larva.

Larvae of *Pentaneura* sp., another dipteran insect, play a central role in Cone Spring feeding dynamics, being present in large numbers in every life stage, throughout the year. They feed on other animals (including other *Pentaneura*), but can survive to the adult stage with nothing but detritus as food. The estimation of production for this species provides a good example of the problems involved in field studies and we will consider it in some detail. In *Pentaneura*, production includes moulting, emigration (emergence of the adult and its subsequent flight) and mortality (consumption by higher trophic levels). Production was calculated using an adaptation of Ricker's (1946) formula for estimating the production of fish populations:

$$Pt = Poe^{(k-i)t} \tag{9.1}$$

where Pt and Po are the energy contents of populations (in joules) present at the end and the beginning of a time interval, k is the instantaneous growth rate (see chapter 6), i is the instantaneous mortality rate and t is the time interval, in this case a month.

The instantaneous growth rate k was estimated from field data and substituted in Ricker's equation, along with population values for each month, to derive a value for i. This instantaneous mortality was multiplied by the mean population of that particular month to yield total mortality for that month. Successive monthly mortality estimates were summed to yield a total annual mortality with an energy content of 0.403 MJ m^{-2} yr^{-1}. This is an estimate of annual production only if there was no change in standing crop over the year. In this population, however, cannibalism occurred, so part of the production was recycled. Therefore mortality losses were adjusted by an amount equal to the energy withdrawn from the population by itself. Figures derived from Ricker's formula are not true

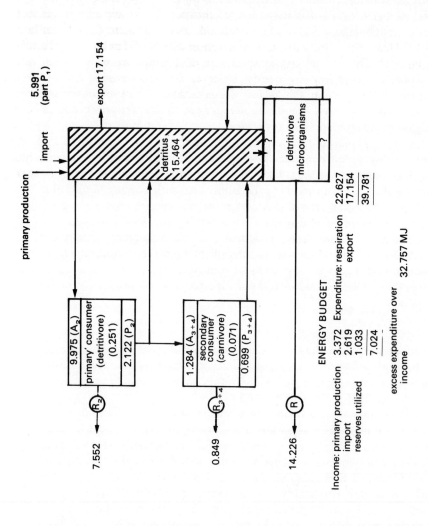

primary production

import

5.991
(part P₁)

export 17.154

detritus
15.464

detritivore
microorganisms

9.975 (A₂)
primary consumer
(detritivore)
(0.251)
2.122 (P₂)

1.284 (A₃₊₄)
secondary
consumer
(carnivore)
(0.071)
0.699 (P₃₊₄)

R₂   7.552

R₃₊₄   0.849

R   14.226

ENERGY BUDGET

Income: primary production   3.372
        import   2.619
        reserves utilized   1.033
                              ─────
                              7.024

Expenditure: respiration   22.627
             export   17.154
                            ──────
                            39.781

excess expenditure over
income   32.757 MJ

Figure 9.6   The energy budget for Cone Spring. (After Tilly, 1968.) Units MJ m⁻² yr⁻¹. Figures in parentheses are standing crop.

mortality losses, however, because emigration and moulting are included. Emigration accounted for $0.025 \, MJ \, m^{-2} \, yr^{-1}$ of the net annual loss leaving a true mortality of $0.378 \, MJ \, m^{-2} \, yr^{-1}$. Cannibalism accounted for 35 %, or $0.132 \, MJ \, m^{-2} \, yr^{-1}$, of this. Only 30 % of this value was actually "reconsumed", the remainder being passed to other trophic levels. Therefore the amount of energy taken from *Pentaneura* production by cannibalism was $0.040 \, MJ \, m^{-2} \, yr^{-1}$ and this was not passed to other trophic levels. Tilly noted that the standing crop of *Pentaneura* diminished by $1647 \, J \, m^{-2}$ during the year and this would have appeared in the Ricker production estimate. It is production from previous years, however, which just happened to be expended in this year, so it is not part of production for the year in question. Therefore this amount and the production reconsumed by the population must be subtracted from Ricker's production estimate before a true value for population production is achieved. Thus production by *Pentaneura* was $0.403 - (0.040 + 0.00167)$ or $0.361 \, MJ \, m^{-2} \, yr^{-1}$, $0.289 \, MJ \, m^{-2} \, yr^{-1}$ being derived from its predatory activities and $0.072 \, MJ \, m^{-2} \, yr^{-1}$ from consumption of detritus.

Three other species were major carnivores of which the planarian *Phagocata velata* was the most important. This species assimilated $0.527 \, MJ \, m^{-2} \, yr^{-1}$ of which $0.205 \, MJ \, m^{-2} \, yr^{-1}$ was obtained from *Pentaneura*. Primary consumer production available to the remaining carnivores was about $0.950 \, MJ \, m^{-2} \, yr^{-1}$ but they required about $1.096 \, MJ \, m^{-2} \, yr^{-1}$, leaving a deficit of $0.146 \, MJ \, m^{-2} \, yr^{-1}$. This imbalance was regarded as further evidence that *Pentaneura* assimilates a significant amount of energy from the detritus store.

The Cone Spring community energy budget is summarized in figure 9.6. Energy expenditure must have been at least $39.781 \, MJ \, m^{-2}$ during the year (consumer respiration + export + $1.033 \, MJ \, m^{-2}$ decline in standing crop). Therefore about $38.748 \, MJ \, m^{-2}$ must have been produced. The harvesting methods used to measure Cone Spring primary production gave an estimate of only $3.372 \, MJ \, m^{-2}$ and the debris yielded a further $2.619 \, MJ \, m^{-2}$, a total income of only $5.991 \, MJ \, m^{-2}$. The discrepancy of about 33 MJ is unexplained, but Tilly believes that income was grossly under-estimated and that the estimate of expenditure was about right. This vividly illustrates a serious problem with ecosystem studies—if an imbalance results it is difficult to check where the errors lay because of the complexity of the study. We cannot be sure whether a series of small errors was compounded when calculating expenditure or whether, as Tilly believes, one large error in calculating income was responsible for the discrepancy. Whatever the causes, we must remember the likelihood of error and treat ecosystem energy calculations with caution. The respectability afforded to ecological ideas by a cloak of numeration may often be illusory!

Using the information of figure 9.6, in which all carnivores are amalgamated (P_{3+4}), we can calculate the ecosystem efficiencies:

$$\frac{P_2}{P_1} = 5\%$$

$$\frac{P_{3+4}}{P_2} = 33\%$$

$$\frac{A_{3+4}}{A_2} = 13\%$$

assuming $P_1 = 39.78 \, \text{MJ m}^{-2}$ (total energy expenditure, and not estimated income).

The considerable discrepancy between P_{3+4}/P_2 and A_{3+4}/A_2 is due to large differences in the P/A ratios for primary and secondary consumers, 21% for detritivores and 54% for carnivores. If we recall the population energy efficiencies shown in table 8.3, aquatic poikilotherm detritivores have a mean P/A value of 32% and the mean figure for all poikilotherm carnivores is 36%. The high P/A ratios for the carnivores at Cone Spring are unexpected and unexplained, and suggest that the ecosystem energy ratios calculated for Cone Spring should be treated with reserve.

This study of energy flow through a simple ecosystem illustrates both the potential and the weakness of ecosystem energy studies. We should have discerned energy flow—its major pathways, their relative efficiencies and their interactions—giving real insight into the system. In fact we have a picture beset with questions. Why did the energy budget not balance? What was the true role of *Pentaneura*? Were the assumptions and techniques valid? Why did major trophic groups, such as carnivores, have quite different P/A ratios than those we would have predicted? Unfortunately, we do not know the answers to any of these important questions.

9.5　The transfer of energy between trophic levels

Most ecosystems show a dynamic but steady state whereby all production is eventually consumed or transported out of the system. The consumption of production by the next trophic level can be expressed by the ratio $Cn/Pn - 1$ (figure 9.3 VI) and unless population biomass increases or some production remains undecomposed it must account for all production (P).

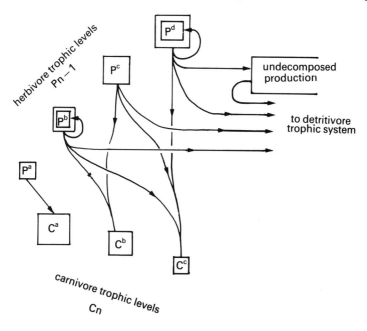

Figure 9.7 The relationship between herbivore production (P^a–d) and carnivore consumption (C^a–c). See text for explanation.

The transfer of energy between trophic levels is the basis of food chains and webs and is an important feature of an ecosystem. The basis of the ratio $Cn/Pn-1$ is the feeding activity of individual populations when confronted with differing amounts and quality of food. We often have little idea of how animals feed (how much, or what, or when they eat). The various possibilities are depicted diagrammatically in figure 9.7. The producers of a particular trophic level, in this example herbivores, comprise numerous species and are exploited in different ways. P^a is eaten solely by the consumer C^a, which is part of the carnivore trophic level. No component of P^a dies naturally or is eaten by another organism. P^b contributes to increase in the biomass of the population, that is it remains alive and uneaten (depicted by an increase in the size of the box), to two carnivore populations (C^b and C^c) and to the detritivore trophic system.

Information and insight into the factors controlling $Cn/Pn-1$ come from three sources: from ecosystem studies which fully describe the utilization of P by any trophic level, from data obtained in studies of population energetics, and from information on feeding. Classical early works, such as those of Ivlev (1939) on the mud-dwelling worm *Tubifex tubifex*, and Gauld (1951) on marine filter-feeding copepods, throw light on the biological basis of energetic ratios.

Ricklefs (1980) has compiled a list of values for $Cn/Pn-1$ (which he terms *exploitation efficiency*) for various animal populations. The value of $Pn-1$ is often not deducible from individual population studies and Ricklefs calculates his exploitation efficiency using an estimate of the food (or production) available for consumption. This is not surprising, because the sort of population energy study normally undertaken would need to be extended considerably to estimate the total production of all possible food sources for the animal population under investigation. Ricklefs' list of exploitation efficiencies, which are really ratios of food consumed (Cn) to food available (part of $Pn-1$), are given in table 9.2. Fifteen out of sixteen of his consumer populations are herbivores and the exploitation efficiencies for this trophic level vary from 1.6 % to 92 % or more. Animals showing the lowest values are nearly all herbivores feeding on general vegetation whereas the more specialized seed feeders, for instance, exploit the available food more effectively. The most striking feature, however, is the extremely wide variation, which prevents us making generalizations about the efficiency of energy transfer from one trophic level to the next, but most

Table 9.2 Exploitation efficiencies of consumer populations (from Ricklefs, 1980) in rank order

Species	Cn/part Pn − 1 (available food)	Food source
vole	1.6 %	vegetable food
cottontail rabbit	2.5 %	vegetation
grasshopper	2.7 %	vegetation
saltmarsh grasshopper	3.0 %	vegetation
white-tailed deer	4.5 %	vegetation
leafhopper	4.6 %	plant sap
beech forest rodents	5.8 %	seeds
African elephant	9.6 %	vegetation
limpet (*Ferrisia rirubsis*)	10.4 to 31.2 %	algae
field mouse	10 to 50 %	seeds
savanna sparrow	10 to 50 %	seeds
least weasel	31 %	mice
midge (*Calospectra dives*)	50 %	detritus
kangaroo rat	86.5 %	seeds
snail (*Tegula funebralis*)	92 %	algae
harvester ant	up to 100 %	seeds

individual populations probably consume only a small fraction of available food resources.

Slobodkin (1959, 1960) devised a laboratory ecosystem of three species. The primary producer was the unicellular alga *Chlamydomonas reinhardi*, the primary consumer (herbivore) was the small crustacean *Daphnia pulex* and the secondary consumer (carnivore) was Slobodkin himself. He aimed to study energy transfer between the three trophic levels when there were different amounts of food available to *Daphnia* and when it was subjected to different predation regimes. Different amounts of food were given to five distinct *Daphnia* populations every 4 days and predation occurred at the

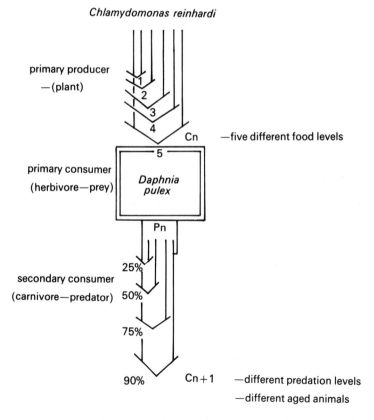

Figure 9.8 Energetic studies by Slobodkin (1959, 1960) on *Daphnia pulex* and *Chlamydomonas reinhardi*.

same time. The numbers of newborn were estimated every 4 days and either 50% or 90% of them removed (young removal). In some experiments similar fractions of large animals were removed instead and in these cases additional predation levels of 25% and 75% were included (adult removal). The experiments are summarized in figure 9.8—further experimental details are given by Slobodkin (1959). They allow the energy equivalent of *Chlamydomonas* consumed per unit time by *Daphnia* (Cn) and the energy equivalent of *Daphnia* consumed per unit time by the predator (Cn + 1) to be assessed.

The effectiveness with which the food supply of *Daphnia* (*Chlamydomonas*) is exploited by the predator (Slobodkin) is given by the food chain efficiency. This equals

$$\frac{\text{the energy equivalent of prey consumed by predator (Cn + 1)}}{\text{the energy equivalent of food supplied to prey}} \qquad (9.2)$$

If all available food is eaten then the food chain efficiency is the same as Cn + 1/Cn and the same controlling factors set its value. These are the A/C and P/A ratios for the populations. Factors likely to control A/C and P/A for individuals have been discussed in chapter 8 and include the quality of food and the size of the respiration component of the budget. Food chain efficiency may be lower than its theoretical maximum for two reasons:

(1) All food (*Chlamydomonas*) is not eaten—too few animals (*Daphnia*) to eat it (predation by Slobodkin too severe). All food is not eaten—animals present are too small to utilize it all (predation strategy is poor—selecting wrong animals).

(2) All food may be eaten but the animals eating it have a low A/C and/or P/A ratio. A/C and P/A may change with the age of the animal. Adult animals past the age of reproduction may have a low P and a high R, for instance.

When Slobodkin removed mainly adult animals, leaving a preponderance of young small ones, food chain efficiency rose to about 10% at a low food level and moderate predation but was reduced at more abundant food levels and high predation (figure 9.9a). Those *Daphnia* remaining after predation were not sufficiently numerous or large to consume all the algae supplied. When young animals were selectively removed (figure 9.9b) food chain efficiency at different food levels became similar, indicating that all available food was always utilized, whatever the level of predation. The maximum food chain efficiency achieved, however, was considerably lower than when young animals were dominating the population, and this

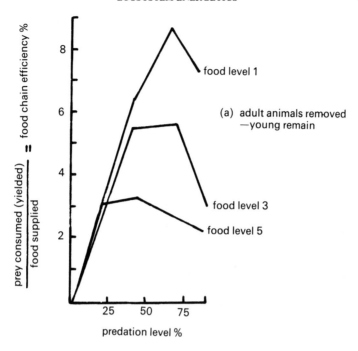

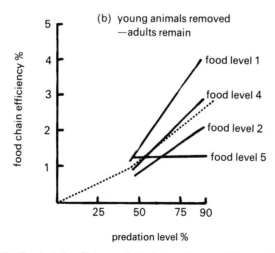

Figure 9.9 Food chain efficiency of *Daphnia pulex* populations at different food levels and different predation levels. (After Slobodkin, 1959.)

indicates that although all food was eaten the *Daphnia* were inefficient at converting it to production which could then be exploited by the predator.

Slobodkin did not calculate the ratio $Cn/Pn-1$ but his experiments show that the way in which predation occurs can affect the ratio quite fundamentally, because predation influences $Cn-1$, A/C, P/A and hence $Pn-1$.

9.6   Changes in ecological energy efficiencies at different trophic levels

Both Lindeman (1942) and E. P. Odum (1971) concluded that the efficiency of energy transfer, both within individual animals and between communities at different trophic levels, increases at higher trophic levels. In 1968, Kozlovsky used available information from community energy studies: from Lake Mendota, USA (Juday, 1940); a temperate cold spring, USA (Teal, 1957); Silver Springs, Florida, USA (Odum, 1957); Cedar Bog Lake, USA (Lindeman, 1942) and a salt marsh, Georgia, USA (Teal, 1962), to test the validity of the statements of Lindeman and Odum. Kozlovsky plotted or recalculated 14 ecological energy efficiencies, which included all the energy efficiencies presented in this book, namely P/C, A/C and P/A for individual animals or populations (chapter 8) and C_3/C_2, A_3/A_2, P_3/P_2 and C_3/P_2 introduced in this chapter (figure 9.3). Kozlovsky came to several "interesting conclusions". With respect to energy transfer within individuals or populations from trophic level 4 (secondary consumers) P/C rises then falls; A/C rises; P/A falls. When energy is transferred between trophic levels 1 and 2, 2 and 3, 3 and 4 (only efficiencies between trophic level 2 and 3 are quoted here) C_3/C_2 increases then decreases; A_3/A_2 is constant at 10%; P_3/P_2 is practically constant at 10%; C_3/P_2 increases and is then constant. These trends are shown in figure 9.10 and Kozlovsky concludes that the generalizations of Lindeman (1942) and E. P. Odum (1971) are not supported by his own analysis.

This chapter has not considered the energy relations of plants and consequently we cannot comment on the relative efficiencies of trophic level 1 (plants) and 2 (herbivores) but the contention that A/C rises from trophic level 2 to 3 (carnivores) and P/A falls is in line with what we would expect from studying individual animals. The decline of P/C from trophic level 2 to 3 suggested by ecosystem studies does not occur either for laboratory animals or populations living in the field, however. Poikilotherms show a distinct rise in the P/C from the herbivore level to the carnivore and the sparse data for homeotherms suggest little change (see table 8.3). Most ecologists have expressed considerable interest in the

(a) individuals and populations

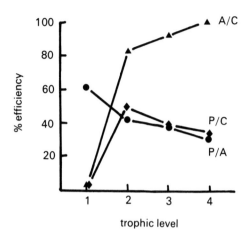

(b) between trophic levels

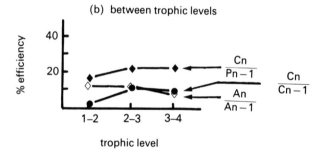

Figure 9.10 Energy transfer (*a*) within individuals and populations and (*b*) between trophic levels. (After Kozlovsky, 1968.)

"between trophic level efficiencies" ever since Lindeman (1942) suggested they might represent "a fundamental trophic principle". We have seen that Slobodkin's (1959, 1960) experiments with *Daphnia pulex* suggested that the maximum efficiency of the conversion of food into biomass yielded to the next trophic level (C_3/C_2) was 12.5% (see figure 9.9*a*). However, further studies, involving Hydrida (Slobodkin, 1964) led him to propose a more or less constant C_3/C_2 of around 10%. Subsequent empirical data eventually convinced Slobodkin (1970) that there was really no upper limit (or indeed anything constant) about such an efficiency, when calculated, from one environmental situation to the next. He concluded that because population consumption is a variable outside the immediate control of the population itself, normal evolutionary selection processes cannot affect it. Therefore C_3/C_2 is a ratio of two global variables and immune to the maximizing and minimizing effects of natural selection (Slobodkin, 1972).

Although the ecological ratios quoted by authors are often poorly defined (and when they are, C_3/C_2, P_3/P_2 and A_3/A_2 are usually regarded as the same) there is still a general feeling that a figure of 10 % is somehow correct. Colinvaux (1972) states that "10 % for animal levels is commonly, though cautiously used", while Krebs (1978) suggests that "Lindeman's efficiency appears to be a constant around 10 % for each set of trophic levels" but he goes on to comment that this may be wrong as figures of 70 % have been recorded for marine food chains. Kucera (1978) states that "ecosystem (or inter-trophic) efficiency is 10 %, a convenient, much quoted value" but then goes on to suggest it is probably too high for most ecosystems. Clapham (1973) tells us that Kozlovsky's 10 % figure indicates that we can expect a 90 % loss of energy between trophic levels and that this seems "fully warranted". Whittaker (1975) and E. P. Odum (1971) still believe that these efficiencies increase at higher trophic levels, Odum quoting 10–20 % between secondary trophic levels and Whittaker from 15 % downwards between trophic level 2 and 3. In one of the early summaries of ecological animal energetics, Phillipson (1966) suggested that values ranged from 5 to 30 % for natural ecosystems, but that more data were necessary before a final decision about the constancy of C_3/C_2 could be made. It is clear that this is still the situation, and the assumption that has grown up that between-trophic-level efficiencies are constant and a numerically convenient 10 % is unfounded, based more on the desire and need for simple understandable ecological generalizations than on fact. Good enough for a rule-of-thumb—but not good enough for a computer model!

The increase in real understanding of how populations and ecosystems work, as a consequence of energy studies, has been disappointing. Between 1968 and 1975 the International Biological Programme, an international co-operative study of ecosystem structure and function, formulated various models which would permit the prediction of secondary production both within trophic levels and between ecosystems. Various A/C and P/A ratios for taxonomic and trophic categories were collated by Heal and Maclean (1975) and used in a grassland ecosystem trophic model which would enable heterotroph production to be calculated where values of net primary production and heterotroph production were already known. The values predicted by the model were generally within an order of magnitude of observed levels, with herbivores accounting for 1.6% of heterotroph production and bacteria and fungi 94% of total production. They concluded that the model provided a mechanism for the prediction of production in ecosystems and that it would enable the effects of variation in composition and energy efficiency of different organism groups to be elucidated. Heal and Maclean (1975) were also able to make significant statements concerning the roles of herbivores and detritivores in ecosystems. Because herbivores eat living plants, events which affect their abundance will interact with and modify the capture of light energy by green plants. Herbivore activity can cause overgrazing and this may destroy large parts of the photosynthetic machine with obvious consequences for all living organisms. Detritivores, on the other hand, are saprophytic and do not directly influence the rate at which energy enters the ecosystem. Energy entering the detritivore feeding system is lost only through respiration and is recycled within the system. These features, together with the large amount of plant production which finds its way into the detritivore system, allow a greater number of links in the food chain, a greater trophic complexity and a greater standing crop of detritivore-based organisms than are found in herbivore food chains.

Ecosystems are very complex and subject to a wide variety of controlling factors, many of which are not easily elucidated in energy terms. With hindsight it is not surprising that simple quantitative generalizations concerning ecosystem energy flow have not been forthcoming, but it is very encouraging that comparative energy studies of ecosystem components, using trophic models, are yielding significant advances in ecological understanding.

SUMMARY

1. Eltonian pyramids are not adequate models of ecosystem trophic relationships because they do not depict the presence and activity of detritivores, which are responsible for decomposition. In woodland ecosystems, for instance, more than 90% of primary production may be consumed by detritivores and less than 10% by herbivores.

2. A more comprehensive model of ecosystem energy flow includes autotrophs (plants), primary consumers (herbivores and detritivores), secondary and tertiary consumers (carnivores) and omnivores. Populations and organisms do not always fit easily and sensibly into a particular trophic level, however.

3. Because all living organisms require energy we should be able to predict the size and number of trophic levels in an ecosystem by studying energy efficiencies. Three efficiencies,

A_3/A_2, P_3/P_2 and C_3/C_2 can be calculated to indicate energy flow from one trophic level (2) to the next trophic level (3) and a further efficiency, C_3/P_1, deals with energy flow between three trophic levels. There is a widely held belief that A_3/A_2 (the so-called Lindeman efficiency) is preferable and ecologically more meaningful. Many authors regard A_3/A_2, P_3/P_2 and C_3/C_2 as exactly comparable, which they are not.

4. Energy flow in Cedar Bog Lake was studied by Lindeman, who concluded that consumers at progressively higher trophic levels in the food cycle were more efficient in the use of their food energy. He also thought that typical ecological efficiencies of an animal trophic level were about 10%.

5. Energy flow in another American freshwater ecosystem, Cone Spring, Iowa, was investigated by Tilly. Primary consumers were detritivores rather than herbivores and 10 species of animals were responsible for 90% of the energy flow. Energy expenditure was at least $39\,\text{MJ}\,\text{m}^{-2}\,\text{yr}^{-1}$ but estimates of primary production and debris import were only $6\,\text{MJ}\,\text{m}^{-2}\,\text{yr}^{-1}$. The discrepancy is unexplained but almost certainly lies in a gross under-estimate of primary production.

6. The transfer of energy between trophic levels is governed by the ratio $Cn/Pn-1$ (often called exploitation efficiency) and its basis is the feeding activity of individual populations when confronted with differing amounts and quality of food. Exploitation efficiencies range from 1.6% to over 90% and this wide variation makes generalization impossible.

7. Experiments performed on *Daphnia pulex* by Slobodkin suggested that the maximum efficiency of the conversion of food into animal biomass yield to the next trophic level (C_3/C_2) was 12.5%, and this led to the proposal of a more or less constant C_3/C_2 ratio of around 10%. Additional evidence, however, suggests that there is no upper limit or indeed anything constant about this efficiency.

CHAPTER TEN

ENERGY AND FOOD PRODUCTION

Energy enters the earth's ecosystems as solar energy, as heat flowing from within the earth and as tidal energy. Solar radiation is about 7700 times greater than the other sources combined and is therefore responsible for supplying most of the energy for the earth's budget. The flow of energy to and from the earth is shown in figure· 10.1. We see that 30% of solar radiation is directly reflected back into space from clouds, water and land surface, while about 47% is absorbed by the atmosphere, by water and by the land, providing the temperature range within which life exists. Much of the remaining 23% powers the movements of water and air and generates the water cycle which is essential to terrestrial life. Only 0.0225% of incoming solar radiation is absorbed by the leaves of plants and used in photosynthesis, eventually providing all of the food energy for animals. During the last few hundred years the earth's ecosystems have become progressively dominated by one species of animal, mankind, and there are few places in the world which escape his influence. In the 1980's there will be over 4000 million people inhabiting the world. This chapter considers the energetics of the production of their food.

A little over 100 years ago sunlight provided almost all the energy needs of mankind but now only about 10% of the world's energy consumption comes from this renewable source and food production utilizes most of the 10%. The fruits, nuts and game of natural ecosystems can support only about 200 million people and today's population is fed by agricultural technology, which has replaced much of the world's natural vegetation with crops or pastures. Estimates of the extent and quality of natural ecosystem replacement by agriculture vary considerably. H. T. Odum (1971) suggests that 10% of the land area of the world is cultivated and a further 31% is grassland and pasture, much of which is managed. The cultivated land produces 13% of the total gross primary production for terrestrial ecosystems while the grassland and pasture accounts for a

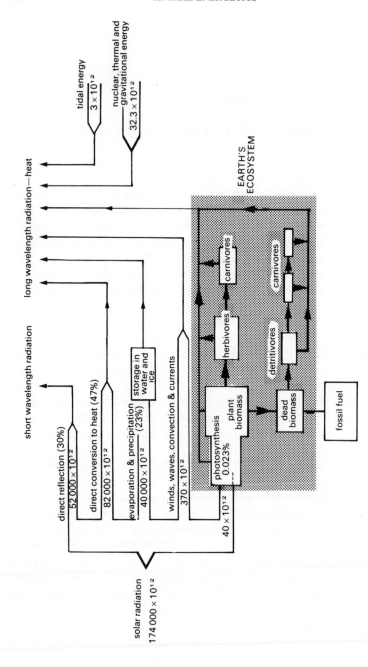

Figure 10.1   Energy flow of the earth in watts (J sec^{-1}). (Adapted from Cook, 1976.)

further 18%. According to Odum, 40% of cultivated land is managed or exploited using a technology driven by non-renewable energy, and this can produce 50 times as much protein per hectare as could be harvested from a natural ecosystem (Cook, 1976). The developed and affluent countries of the world rely almost entirely on this type of ecosystem for their food. The poorer countries, where the population is often undernourished, are trying to develop such a system. Intensive agriculture and the intensive exploitation of natural resources require an input of energy (in the form of ploughing, sowing, harrowing, reaping) which is derived from non-renewable fossil fuels. This, in effect, supplements the energy derived from sunlight. It is outside our scope here to review comprehensively the energetics of world agriculture, so we shall concentrate on the intensive agricultural systems. It is only these which can feed the world's human population adequately. There are many facets to energy use in this context: grain and live stock production, fruit and vegetable production, fisheries, food processing, packaging and transport. Pimentel and Pimentel (1979), Leach (1976) and Schneider (1976) discuss many of these aspects in depth.

10.1 Energy and nutritional requirements of man

Man is an omnivore but social, economic and political forces determine specific diets in various parts of the world. They range from those composed primarily of plant material to those where animal products are both valued and desirable. Human populations in some places are continually threatened with starvation, while others overeat. In some areas population and hence food requirement is increasing while in others it is stationary or declining.

A comprehensive survey of world food energy and protein requirements is not feasible so we will use specific examples to illustrate general principles. The majority of mankind eats plant material to obtain energy and other nutrients, consuming on average 8790 kJ day^{-1}, mainly derived from grains and legumes. In Central America a labourer eats about 500 g of maize and 100 g of black beans a day and these provide 8870 kJ and 68 g of protein. The maize and beans provide the essential amino acids, and when possible additional energy and nutrients are obtained from other plant and animal products. People in the USA, however, have a daily protein intake of 101 g of which 69 g are of animal origin and this is fairly typical of highly industrialized nations (Pimentel and Pimentel, 1979). Although the quantity of protein obtained from 85 g of cooked ground beef is similar to that in 190 g of cooked dried beans, the nutritional quality is quite different. Animal proteins contain the eight essential amino acids for man in optimum amounts whereas plant proteins contain less of some essential amino acids or are deficient in one or more. Cereal grains for instance contain relatively low amounts of lysine, while legumes (peas and beans) are relatively low in methionine but have ample lysine. Often it is possible to complement plant foods to provide a balanced amino acid supply, but the notion remains that animal protein is of "high quality" and is therefore desirable. Although it is possible to devise a diet which is nutritionally acceptable and devoid of animal protein, pregnant women and growing children find it difficult to consume the quantity of plant material necessary to provide enough essential nutrients such as calcium and iron, while vitamin B_{12} is entirely lacking from plant food. Also, animal products contain more energy per unit weight than do plants (1620 kJ can be obtained by eating 455 g of sweet corn or 140 g of beef) and this is an important factor in the nutrition of children. There is thus a nutritional requirement in man for both plant and animal food products overriding any sociological, ethnic or economic factors. In the remainder of this chapter we shall consider the energetics of grain and milk production, the exploitation of marine fisheries and conclude with a general summary of energy use in intensive agriculture.

10.2　The energetics of grain production

Plants provide over 70% of the protein consumed by man, and are also consumed by livestock, which supply much of the remainder. About 90% of the plant protein utilized directly by man is derived from 15 major crops and of these the cereal grains are the most important, yielding almost 50% of the protein available to mankind worldwide (Pimentel and Pimentel, 1979). Cereals can be cultivated in a wide variety of soil types, different rainfalls and temperatures and they yield large quantities of nutrients per unit of land area. They have a relatively low moisture content at harvest (15%, compared to 80% in potatoes) and this allows more efficient transport and easier storage.

Leach (1976) gives several energy budgets for wheat grown in the UK between 1968 and 1972. Two energy budgets are for winter wheat (seed sown in the autumn and overwintering in the soil) and two are for spring wheat. Much of the high productivity achieved by wheat is due to fertilizer application. Fertilizers are costly and so Leach chose four different fertilizer regimes when constructing his energy budgets, which are shown

Table 10.1 The energetics of wheat production in the UK under 4 different fertilizer treatments (after Leach, 1976)

	Spring sown		Winter sown	
Fertilizer input ($kg\,ha^{-1}\,yr^{-1}$)				
Nitrogen	97	75	95	130
Phosphorus	48	38	55	50
Potassium	48	38	55	50
Total	193	151	205	230
Energy input ($GJ\,(=10^9\,J)\,ha^{-1}\,yr^{-1}$)				
Fertilizer	8.86	6.86	8.87	11.67
Tractor and equipment	4.53	4.53	4.53	4.53
Sprays (4 kg)	0.40	0.40	0.40	0.40
Drying fuel etc.	1.76	1.96	2.23	2.29
Total	15.55	13.75	16.03	18.89
Output ($tonnes\,(t)\,ha^{-1}\,yr^{-1}$)				
Net seed yield	3.20	3.58	4.08	4.21
Ratio				
Energy input per t grain ($GJ\,t^{-1}$)	4.86	3.84	3.93	4.87

Average for UK wheat
Energy input   = 17.8
Net yield grain ($t\,ha^{-1}\,yr^{-1}$) = 3.9
Energy output in grain ($GJ\,ha^{-1}\,yr^{-1}$) = 56.2
Protein output in grain ($kg\,protein\,ha^{-1}\,yr^{-1}$) = 400
Energy out/energy in (Er)   = 3.35

in table 10.1. The more fertilizer applied, the greater the yield, but a 32% increase in fertilizer (expressed in energy terms) applied to winter wheat leads to an increase in seed yield of only 0.13 metric tonne (t) or 3%. Although any increase in yield is desirable, the energy cost of this small rise is out of all proportion and leads to an unfavourable ratio of energy input to grain yield when compared with the other winter-sown wheats.

These energy budgets are for specific crops grown in particular places, and when Leach compiled data on a countrywide basis he found that variations from farm to farm (no doubt due to different soil types, climate

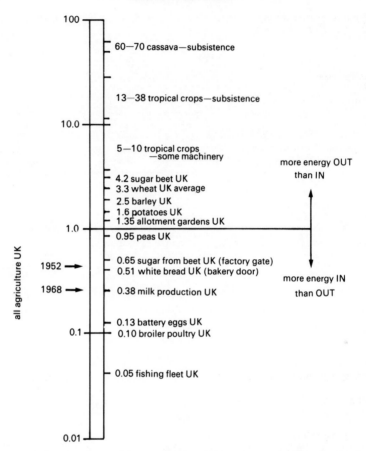

Figure 10.2 Energy ratios (Er) for different food crops and different agricultural systems. (After Leach, 1976.)

and cultivation techniques) were very high. The variations in input and output were much wider than those seen in table 10.1, and the average energy budget for UK wheat shown in table 10.1 may therefore be inaccurate. The ratio of energy output to energy input calculated for all wheat grown in the UK is 3.35. This figure, the *energy ratio*, or Er, (Leach, 1976) for wheat, only has meaning when compared with other such ratios; figure 10.2 shows Er values for a variety of food production methods worldwide. We shall use it as a yardstick when considering the energetics of food production. It is worth stressing that the higher the Er value the greater the energy output for a given energy input. If the Er value falls below 1 the energy expended in food production is greater than that obtained from the food. The high Er value for wheat grown in Britain indicates that it is one of the more energy-efficient food production systems.

Energy ratios for wheat grown intensively in the UK, USA and India (using bullocks and no fertilizer) are given in table 10.2. The Er for Indian wheat is over 60, with the only energy input being that expended in machinery manufacture (because yields in this table are net—the seed used to grow the crop is deducted from the yield and therefore does not show as an input). Bullocks and man provide the power. In intensive wheat-

Table 10.2 A comparison of the energetics of wheat production in the UK (Leach, 1975), USA and India (Pimentel and Pimentel, 1979)

	USA (average)	UK (winter wheat)	India (non-mechanized)
Energy in $(GJ\,ha^{-1}\,yr^{-1})$			
Fertilizer	3.53	8.87	0
Cultivation and processing	5.29	7.16	0.17
Total	9.82	16.03	0.17
Energy out $(GJ\,ha^{-1}\,yr^{-1})$			
Net grain yield	27.03	58.75	10.48
Energy ratio			
Er	2.75	3.66	60.21
Fertilizer and processing analysis			
Total amount fertilizer $(kg\,ha^{-1}\,yr^{-1})$	106	205	0
Energy content of fertilizer $(GJ\,kg^{-1}\,yr^{-1})$	3.53	8.87	0
Energy cost of fertilizer $(GJ\,kg^{-1})$	0.033	0.043	0
kg grain produced per kg fertilizer	19.43	18.08	
kg grain produced per GJ cultivation and processing energy	5.11	8.21	

growing systems the energy input from man is insignificant (about 0.1 % of total input in US wheat, for instance) and for that reason it is omitted from the energy budgets. The Indian wheat is labour-intensive whereas British and American wheat growing is fossil fuel-intensive. The labour energy input is one-hundredfold greater in the former case and consequently is a significant factor. When it is added to the energy input by bullocks, the Er for Indian wheat falls to 0.096 and more energy is now being expended than gained. In effect wheat cultivation in India is being "energy subsidized" by the photosynthesis of the grasses used to feed the bullocks. Although viewed in this way Indian wheat cultivation is very inefficient, mainly because the yield is so low, it does utilize a renewable energy source (grass) for which there would be little other use.

With an energy ratio of 3.66, British winter wheat production is more energy efficient than its American counterpart, because different amounts of fertilizer are used. British wheat receives almost double the weight of fertilizer given to American wheat and when the energy cost of fertilizer production is taken into account it is over twice as expensive: 8.87 GJ per hectare (1 GJ equals 10^9 J) compared with 3.53 GJ ha^{-1} (table 10.2). Wheat yields are correspondingly higher in Britain but very similar quantities of grain are produced for equivalent amounts of fertilizer (table 10.2). In fact, the better energy efficiency of British wheat is not accounted for in terms of greater yield per unit of energy used in fertilizer manufacture. The real reason for the difference in these energy ratios lies in cultivation and processing inputs, which do not increase in the same proportion as yield increases. American wheat yields 5.11 kg grain per GJ expended in cultivation and processing, whereas British wheat yields 8.21 kg. This gives the higher energy ratio for British wheat production despite its marginally less efficient use of fertilizer. The increase in yield produced by fertilizers reduces the "fixed unit costs" of cultivation and processing, leading to greater overall efficiency.

The data for wheat cultivation in these three countries provide a good illustration of both the value and the limitations of energy ratios. It is worth considering whether grain output per unit area should be increased in the USA by applying more fertilizer. This would lead to greater input costs, but if yield was increased in the same way as in Britain, then the overall energy efficiency could be raised. Alternatively, the energy cost of cultivation in the USA might possibly be reduced. None of the energy budgets in table 10.2 is completely comprehensive, for they do not include the energy input of sunlight, for instance, and it can be argued that there is no reason to include this renewable energy resource. Indian wheat-growing then appears very efficient. Grain output per unit area is low, however, and satisfies the energy and protein demands of very few people. It is important to remember that energy ratios are only one aspect of food production and in some situations other considerations may override the general desire for energy-efficient agricultural systems. At the very least a clear distinction should be made between the use of renewable and non-renewable energy resources when considering agricultural energy efficiency, as we shall now see.

10.3 The energetics of milk production

Milk and milk products account for 40% by weight of British agricultural output, nearly four times the meat production. Despite its greater bulk, the

Table 10.3 An energy budget for UK milk production by Friesian cows (after Leach, 1976)

Energy in	
(GJ/cow yr)[1]	
Raising heifer	3.66
Grazing and hay	9.48
Concentrated feed	10.62
Straw	0.35
Vet. fees, buildings	1.91
Electricity and water	5.03
Total	31.05
Energy out	
(GJ/cow yr)	
Milk	11.12
Calf (11.3 kg dressed weight)	0.12
1/4 cow[2] (135 kg dressed weight)	0.34
Milk production (t/cow yr)	4.09
Meat production (t/cow yr)	0.15
Milk protein (3.5%) (kg protein)	143.0
Meat protein (13%) (kg protein)	5.86
Energy input/kg milk (MJ)	7.59
Energy ratio E_r	0.374

Notes
[1] A cow year is a 12-month period spanning part of 2 calendar years.
[2] Assumes a cow life of 4 years.

protein and energy content of milk production is similar to that of meat production (Leach, 1976). Dairy farming suits many of the agricultural conditions prevalent in Britain and we shall use it as an example of an animal production system. Dairy farming is also important worldwide because, of all the animal protein products, milk is the most efficient in terms of conversion of plant protein to animal protein (Pimentel and Pimentel, 1979).

Energy budgets for animal production are usually more complex than those for plants because of the more varied nature of the inputs and outputs. Table 10.3 presents a simplified average energy budget for UK milk production based on the input and output of a Friesian cow for one year. The most striking feature of the input is the energy expenditure on feeding the stock with concentrated food and this is a simple and direct measure of the intensity of the farming. There is a clear inverse relationship between energy input and land required for milk production, which indicates that the less land used in the production of, say, 4550 litres (1000 gallons) of milk, the more the energy input required. An additional input of 24 GJ is necessary to reduce pasture by 1 ha and yet keep milk production stable. An equivalent figure for the "intensification" of wheat is 1 GJ (Leach, 1976). A comparison of the energy ratios for wheat (3.35) and milk (1.37), and the trend in figure 10.2, make it clear that animal production, of whatever type, requires considerably greater energy expenditure than plant production. Farm output and consumption of animal products have risen steadily in the past few decades until in 1972 they provided 60% of UK protein intake, compared to a world average of 30%. Even a small reversal of this trend could save substantial amounts of fossil fuel, much of which is used to produce fertilizer to grow the plants to feed to the animals. This would release a lot of land for less intensive farming.

The trend towards high protein output and associated low energy ratios is seen in British agriculture as a whole. If the energy ratios for different types of farm are plotted against their energy output in the form of animal products (figure 10.3) then we see clearly that the greater the percentage of energy in animal products the lower the energy ratio of the whole farm. Thus producing animal protein is less efficient in energy terms than plant production. An interesting point emerges, however, if we consider the average British dairy farm, which has 28% of its energy output and 38% of its protein output in the form of animal products. Its energy input per "total hectare", including land for growing feedstuffs off the farm, is 28.5 GJ ha^{-1}, and energy output is 22.0 GJ ha^{-1}, giving an energy ratio of

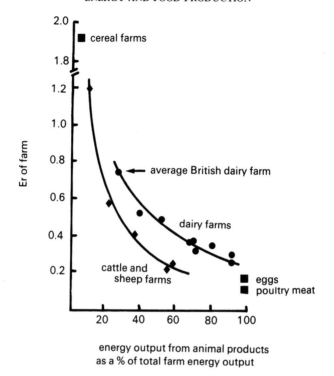

Figure 10.3   The energetics of British farms with particular reference to animal products. (After Leach, 1976.)

0.77. If this pattern could be repeated nationwide, the UK could feed all of its population rather than half, on rather less land than the present area of crops and grass. Furthermore protein output would exceed present consumption by nearly 50%! The energy input for supplying all food requirements in an unprocessed state would be 309 MGJ (1 MGJ equals 10^6 GJ) compared to the 670 MGJ of UK agriculture, fisheries and food imports in 1969 (Leach, 1976).

Because of the complex interdependence between plant production, animal consumption, the availability of land, the cost of fossil fuel, and labour and consumer preference, it is difficult to evaluate the desirability of one type of livestock production against any other. Pimentel *et al.* (1975) give comprehensive comparisons of the energetics of production of various animal products, but it is clear that in a world where human population, and therefore world food demand, are still increasing, and where fossil fuel continues to rise in price, any food production system with a low energy ratio will become more difficult to sustain.

10.4   The energetics of fisheries

As 71 % of the earth's surface is covered by oceans, their potential for supplying food has always been thought important. In 1970 about 70×10^6 t of sea food was harvested from the oceans, most of it bony fish. About 6×10^6 t is directly consumed by man (less than 10 %) whilst about 30 % is fed to livestock which will subsequently be eaten by man. On the basis of these estimates only about 5 % of man's animal protein consumption is met from fish protein (Pimentel and Pimentel, 1979). If 16.75 GJ of light energy are fixed by the phytoplankton of one hectare of ocean in one year, and if there are four links in the food chain (each with a $Pn + 1/Pn$ ratio of 10 %—see chapter 9) before edible fish are produced, then about 1924 ha of ocean are necessary to yield the protein consumption of one person in the USA for one year. This assumes that 40 % of the yield is edible fish. Although this is an unrealistic and extreme situation (very few Americans would be willing to eat only fish to obtain their protein) they do indicate a serious energy problem that is associated with exploiting marine fisheries. Stated simply, only certain marine species are acceptable for human consumption. They have to be searched for, sometimes over long distances and for long periods, and this is an energy intensive process. The energy budget calculated by Leach (1976) for UK fisheries in 1969 is given in table 10.4, and we see that fuel costs comprise a large portion of the energy input, accounting for about 95 % of the total. Not all fisheries are as energy intensive, and data from fisheries off the north-eastern coast of the USA show that an expenditure of 2.2 J of fossil fuel energy is required for each J of human fish protein harvested from inshore waters (9.6 J for offshore fishing). This compares with 20 J of fossil fuel input required to harvest 1 J of fish protein in the UK. There are obviously many complex reasons for this striking difference, but Rochereau and Pimentel (1978) suggest that an important factor in determining the overall energy efficiency of fishing is the size of the fishing vessel: as this increases, energy efficiency declines, because there is a non-linear relationship between vessel size and its gross energy requirement. Twenty-two vessels each of 15 gross registered tons have the same fishing capacity as one vessel of 330 gross registered tons, yet the smaller vessels in total are 44 % more energy efficient in obtaining the same fish yield output. Although large vessels use less labour, the operating costs in terms of a non-renewable

Table 10.4 An energy budget for UK fisheries 1969 (after Leach, 1976)

Energy in (MGJ)	
Fuel	31.83
Equipment	0.74
Ice	0.40
Total	32.97

Energy out (MGJ)	
Edible fish	
—demersal	0.905
—pelagic	0.740
—shellfish	0.020
	1.665
Fishmeal	0.720
Total	2.385
Edible protein (10^3 t)	67.48
Fishmeal protein (10^3 t)	25.48
Energy input/kg edible fish (MJ)	34.6
Energy ratio Er[1]	0.05

[1] For edible fish.

energy source are high and from an energy conservation point of view it would be more sensible and economical to employ more men and use less fuel.

Whatever the present energy economics of fisheries, fish remains a major source of protein and energy for mankind. Unlike agricultural ecosystems, the oceans are largely unmanaged and many fish stocks are either being too heavily fished or hardly exploited at all. If the potential of the oceans to provide protein for human use is to be effectively utilized (and Gulland, 1971, suggests that present catches could be doubled using types of fish already familiar to consumers) then efficient and effective development of fisheries throughout the world is needed. Assessment of the potential of various fish stocks, development of under-utilized ones, rational management of heavily fished ones, and development of new fisheries based on less familiar fish are four aspects which should be

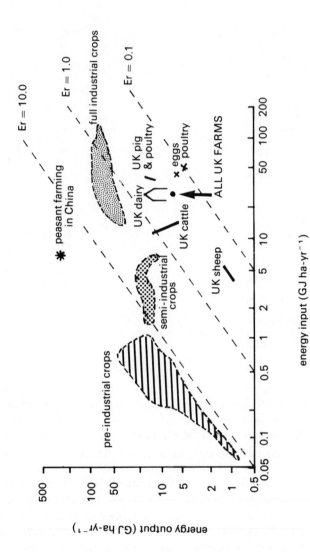

Figure 10.4 Energy inputs and outputs per unit of land for world agriculture. (After Leach, 1976.)

explored. If fisheries are to be an effective and economic protein source in the latter part of the 20th century then attention must be given to the energetics of exploitation. It is not sufficient to conserve and nurture a naturally renewable source of food, for we must also exploit it in an energetically efficient way.

Aquaculture of fish is currently being investigated with a view to the production of high quality protein. In the southern USA, catfish aquaculture occurs on a commercial scale. The total inputs to produce a yield of about 2780 kg of catfish per hectare are 219.7 MJ (1 MJ equals 10^6 J) of fossil energy. Assuming a 60% dressed weight and 23% protein, 384 kg of catfish protein is yielded, with an energy value of 6.28 MJ. The Er is 0.028 which makes catfish aquaculture significantly less efficient than broiled poultry production (figure 10.2).

10.5 The overall energy efficiency of intensive agriculture

The human population of the world requires both energy and nutrients from its food, and almost all food is provided by agricultural systems which have substantially replaced natural ecosystems in some parts of the world. We have investigated the energetics of three important, but quite different, food-producing systems associated with the developed world, its affluence and desire for a high protein diet. We have noted and commented upon both large and small differences in energy ratios, but when we consider agriculture worldwide we see an even larger range of energy ratios and energy flows per unit of land area. World agricultural energetics is summarized in figure 10.4, which plots energy output ha^{-1} year^{-1} against energy input ha^{-1} year^{-1}. Although there are specific reasons for many of the anomalies, some broad trends are clear. These remain if protein output is plotted instead of energy output. Farming systems with very low inputs and outputs are shifting agricultures, often operating a 10–17 year cultivation cycle. The hunter-gatherer type of civilization is completely off the left-hand scale of this figure. The !Kung Bushmen of the Kalahari Desert in Africa, for example, have an agricultural output of only 2.9 MJ ha^{-1} yr^{-1} which means that each person requires 10.4 km^{-2} to provide his diet. Outputs for pre-industrial farming (defined as having fossil fuel inputs of less than 10% of the total) rise with energy input, almost all in the form of human or animal work. Chinese peasant farming has the highest energy output, achieved by the cultivation of small plots, intensive manuring and double cropping. During the 20th century British agriculture has been transformed from pre-industrial farming to fully-industrialized agriculture. The latter, intensive, systems have outputs sufficient to support 10–20 people per hectare if they consume an all-vegetable diet, but inputs are also high. There is a distinct trend towards more energy input achieving less increase in output. Since 1945 there has been a rapid development of agricultural mechanization. By 1950 tractors outnumbered horses, and in 1962 tractors outnumbered farm workers. British agriculture has entered the same bracket as heavy engineering in terms of energy use per man employed. In usual economic terms the substitution of fossil energy for manpower, coupled with a considerable increase in output, is highly desirable. But what will happen in the future if, as seems likely, energy costs increase rapidly? There has been a second important trend in British agriculture which has helped to produce diminishing energy returns for a given energy expenditure and that is an increase in the proportion of animal production. Whilst output of food energy rose by 30% between 1954 and 1972, protein output rose by 35%. This provided food for an additional 5.5 million people and resulted in a considerable increase in UK self-sufficiency in food.

The main factor responsible for the low energy efficiency and land-use efficiency of British agriculture (figure 10.4) is the high proportion of food production in the form of animal products. Agriculture is subject to market forces, however, and in Britain consumers like milk, eggs and meat. As an affluent society we can afford to pay for them and will continue to want them. Eating meat is nutritionally and energetically desirable, as about one-third of British farmland is suitable only for pasture and grazing, but it remains to be seen for how long current production and consumption patterns, which have energy budgets heavily dependent on fossil fuel, can be sustained. Soon intensive agricultural energy budgets may be difficult to balance and they have never been relevant to the needs of the developing world, where the greatest challenge to produce food lies.

SUMMARY

1. Solar energy is responsible for supplying most of the earth's energy but only 0.0225% of incoming solar radiation is absorbed by the leaves of green plants and used in photosynthesis, eventually providing all the food energy for animals. 10% of the world's land area is cultivated by man and a further 31% is grassland and pasture. These land areas provide much of the food for mankind.

2. When cultivated land is managed or exploited using a technology driven by non-renewable fossil fuel, 50 times as much protein per hectare is produced as could be harvested from a natural ecosystem. Although the quantity of protein obtained from 85 g

of cooked ground beef is similar to that obtained from 190 g of cooked dried beans, the quality is quite different and there is a nutritional requirement in man for both plant and animal food products.

3. Plants provide directly over 70 % of the protein consumed by man and are also consumed by livestock which supply much of the remainder.

4. The ratio of energy output to energy input for British-grown wheat (the Er value) is 3.35. If the Er value falls below 1 then energy expended in food production is greater than that obtained from the food. The high Er value for wheat grown in Britain indicates that it is one of the more energy-efficient intensive food production systems.

5. Of all the animal protein products milk is the most efficient in terms of conversion of plant protein to animal protein. The more "intensive" the production system (i.e. the less land used directly for the production of a given volume of milk) the more energy input is required. An additional 24 GJ is required to reduce pasture by 1 ha, yet keep milk production stable. The Er value for milk is 1.37, indicating that animal production requires considerably greater energy expenditure than plant production.

6. Animal products provide 60 % of UK protein intake compared to a world average of 30 %. Even a small reversal of this trend could save substantial amounts of fossil fuel. The Er value for an average British dairy farm is 0.77 and if this pattern could be repeated nationwide the UK could feed all its population on rather less than the present area of crops and grass.

7. About 5 % of man's animal protein consumption is met from fish protein but the search for edible and desirable fish is an energy-intensive process. The Er value for British fishing fleets is 0.077, indicating that fish consumption requires the greatest fossil fuel energy expenditure of any food-producing system.

8. The main factor responsible for the low energy efficiency and low land-use efficiency of British agriculture is the high proportion of food production in the form of animal products. It remains to be seen for how long current production and consumption patterns, with energy budgets heavily dependent on fossil fuel, can be sustained.

REFERENCES

Alexander, R. McN. (1977) "Terrestrial locomotion", in *Mechanics and Energetics of Animal Locomotion*, eds. R. McN. Alexander and G. Goldspink, Chapman and Hall, London, pp. 168–203.

Allen, K. R. (1951) The Horokiwi Stream. A study of a trout population. *New Zealand Marine Dept. Fish. Bull.* **10**, 1–231.

Ashworth, A. (1969) Metabolic rates during recovery from protein–calorie malnutrition: the need for a new concept of specific dynamic action. *Nature* **223**, 407–409.

Bartholomew, G. A. and Casey, T. M. (1978) Oxygen consumption of moths during rest, pre-flight warm-up, and flight in relation to body size and wing morphology. *J. exp. Biol.* **76**, 11–25.

Bayne, B. L. and Scullard, C. (1977) An apparent specific dynamic action in *Mytilus edulis* L. *J. mar. biol. Ass. U.K.* **57**, 371–378.

Bernays, E. A. and Simpson, S. J. (1982) Control of food intake. *Adv. Insect Physiol.* **16**, 59–118.

Bertalanffy, L. von (1938) A quantitative theory of organic growth. *Hum. Biol.* **10**, 181–213.

Beukema, J. J. and de Bruin, W. (1979) Calorific values of the soft parts of the tellinid bivalve *Macoma balthica* (L.) as determined by two methods. *J. exp. mar. Biol. Ecol.* **37**, 19–30.

Beverton, R. J. H. and Holt, S. J. (1957) On the dynamics of exploited fish populations. *Fishery Invest., London* (Ser. II) **19**, 1–533.

Blaxter, K. L. (1967) *The Energy Metabolism of Ruminants* (second edition). Hutchinson, London.

Blaxter, K. L. (1971) Methods of measuring the energy metabolism of animals and interpretation of results obtained. *Fedn. Proc. Fedn. Am. Socs. exp. Biol.* **30**, 1436–1443.

Brafield, A. E. and Solomon, D. J. (1972) Oxy-calorific coefficients for animals respiring nitrogenous substrates. *Comp. Biochem. Physiol.* **43A**, 837–841.

Brett, J. R. and Groves, T. D. D. (1979) "Physiological energetics", in *Fish Physiology*, Vol. 8, *Bioenergetics and Growth*, eds. W. S. Hoar, D. J. Randall and J. R. Brett, Academic Press, New York, pp. 279–352.

Brody, S. (1945) *Bioenergetics and Growth*. Reinhold Publishing Corporation (reprinted 1974 by Hafner Press, New York).

Brouwer, E. (1965) "Report of sub-committee on constants and factors", in *Energy Metabolism*, (European Association for Animal Production Publication no. 11) ed. K. L. Blaxter, Academic Press, London and New York, pp. 441–443.

Calow, P. (1976a) "Ecology, evolution and energetics: a study in metabolic adaptation", in *Advances in Ecological Research*, Vol. 10, ed. A. Macfadyen, Academic Press, London, pp. 1–61.

Calow, P. (1976b) *Biological Machines. A Cybernetic Approach to Life.* Edward Arnold, London.

Calow, P. (1977) Conversion efficiencies in heterotrophic organisms. *Biol. Rev.* **52**, 385–409.

Calow, P. (1978) *Life Cycles. An Evolutionary Approach to the Physiology of Reproduction, Development and Ageing.* Chapman and Hall, London.

158

Carey, F. G. (1973) Fishes with warm bodies. *Scient. Am.* **228**, No. 2, 36–44.

Chappuis, P., Pittet, P. and Jéquier, E. (1976) Heat storage regulation in exercise during thermal transients. *J. appl. Physiol.* **40**, 384–392.

Clapham, W. B., Jr. (1973) *Natural Ecosystems.* The Macmillan Co., New York.

Cohen, J. (1977) *Reproduction.* Butterworths, London.

Colinvaux, P. A. (1972) *Introduction to Ecology.* John Wiley and Sons Inc., New York.

Collier, B. D., Cox, G. W., Johnson, A. W. and Miller, P. C. (1973) *Dynamic Ecology.* Prentice-Hall Inc., New Jersey.

Conover, R. J. (1978) "Transformation of organic matter", in *Marine Ecology,* Vol. 4, *Dynamics,* ed. O. Kinne, John Wiley & Sons, Chichester, pp. 221–499.

Cook, E. (1976) *Man, Energy, Society.* W. H. Freeman & Co, San Francisco.

Curtin, N. A. and Woledge, R. C. (1978) Energy changes and muscular contraction. *Physiol. Rev.* **58**, 690–761.

D'Oleire-Oltmanns, W. (1977) "Combustion heat in ecological energetics. What sort of information can be obtained?" in *Applications of Calorimetry in Life Sciences,* eds. I. Lamprecht and B. Schaarschmidt, Walter de Gruyter and Co., Berlin, pp. 315–324.

Elliott, J. M. (1976) The energetics of feeding, metabolism and growth of brown trout (*Salmo trutta* L.) in relation to body weight, water temperature and ration size. *J. Anim. Ecol.* **45**, 923–948.

Elliott, J. M. and Davison, W. (1975) Energy equivalents of oxygen consumption in animal energetics. *Oecologia* **19**, 195–201.

Elton, C. (1927) *Animal Ecology.* Sidgwick and Jackson Ltd. (reprinted 1968 by Methuen Co. Ltd., London).

Engelmann, M. D. (1966) "Energetics, terrestrial field studies and animal productivity", in *Advances in Ecological Research,* Vol. 3, ed. J. B. Cragg, Academic Press, London, pp. 73–115.

Farrell, D. J. (1974) "General principles and assumptions of calorimetry", in *Energy Requirements of Poultry,* eds. T. R. Morris and B. M. Freeman, British Poultry Science Ltd., Edinburgh, pp. 1–24.

Garrow, J. S. (1973) "Specific dynamic action", in *Energy Balance in Man,* ed. M. Apfelbaum, Masson, Paris, pp. 209–218.

Gauld, G. T. (1951) The grazing rates of planktonic copepods. *J. mar. biol. Ass. U.K.* **29**, 695–706.

Gnaiger, E. (1980) Energetics of invertebrate anoxibiosis: direct calorimetry in aquatic oligochaetes. *FEBS Lett.* **112**, 239–242.

Goldspink, G. (1977) "Muscle energetics and animal locomotion"; "Energy cost of locomotion", in *Mechanics and Energetics of Animal Locomotion,* eds. R. McN. Alexander and G. Goldspink, Chapman and Hall, London, pp. 57–81 and 153–167.

Gulland, J. A. (1971) *The Fish Resources of the Ocean.* Fishing News (Books) Ltd., West Byfleet, England.

Hammen, C. S. (1980) Total energy metabolism of marine bivalve mollusks in anaerobic and aerobic states. *Comp. Biochem. Physiol.* **67A**, 617–621.

Harrison, R. and Lunt, G. G. (1980) *Biological Membranes. Their Structure and Function* (second edition), Blackie & Son Ltd., Glasgow.

Heal, O. W. H. and Maclean, S. F. (1975) "Comparative productivity in ecosystems — secondary productivity", in *Unifying Concepts in Ecology,* eds. W. H. van Dobben and R. H. Lowe-McConnell, W. Junk, Hague, pp. 89–108.

Heller, H. C., Crawshaw, L. I. and Hammel, H. T. (1978) The thermostat of vertebrate animals. *Scient. Am.* **239**, no. 2, 88–96.

Hemmingsen, A. M. (1960) Energy metabolism as related to body size and respiratory surfaces, and its evolution. *Rep. Steno meml. Hosp.* no. 2, 7–110.

Hochachka, P. W., Fields, J. and Mustafa, T. (1973) Animal life without oxygen: basic biochemical mechanisms. *Am. Zool.* **13**, 543–555.

Hochachka, P. W. and Somero, G. N. (1976) "Enzyme and metabolic adaptations to low oxygen", in *Adaptation to Environment: Essays on the Physiology of Marine Animals*, ed. R. C. Newell, Butterworths, London and Boston, pp. 279–314.

Hubbell, S. P. (1971) "Of sowbugs and systems: the ecological bioenergetics of a terrestrial isopod", in *Systems Analysis and Simulation in Ecology*, Vol. 1, ed. B. C. Patten, Academic Press, New York, pp. 269–324.

Humphreys, W. F. (1979) Production and respiration in animal populations. *J. Anim. Ecol.* **48**, 427–453.

Ivlev, V. S. (1939) Transformation of energy by aquatic animals. Coefficient of energy consumption by *Tubifex tubifex* (Oligochaeta). *International Revue Hydrobiol.* **38**, 449–458.

Jéquier, E. (1977) "Whole body calorimetry", in *Applications of Calorimetry in Life Sciences*, eds. I. Lamprecht and B. Schaarschmidt, Walter de Gruyter and Co., Berlin, pp. 261–278.

Jobling, M. and Davies, P. S. (1980) Effects of feeding on metabolic rate, and the Specific Dynamic Action in plaice, *Pleuronectes platessa* L. *J. Fish Biol.* **16**, 629–638.

Jones, C. W. (1981) *Biological Energy Conservation* (second edition). Chapman and Hall, London and New York.

Juday, C. (1940) The annual energy budget of an inland lake. *Ecology* **21**, 438–450.

Kammer, A. E. and Heinrich, B. (1978) Insect flight metabolism. *Adv. Insect Physiol.* **13**, 133–228.

Kay, D. G. and Brafield, A. E. (1973) The energy relations of the polychaete *Neanthes* (= *Nereis*) *virens* (Sars). *J. Anim. Ecol.* **42**, 673–692.

Kleiber, M. (1947) Body size and metabolic rate. *Physiol. Rev.* **27**, 511–541.

Kleiber, M. (1950) "Calorimetric measurements", in *Biophysical Research Methods*, ed. F. M. Uber, Interscience Publishers Inc., New York, pp. 175–209.

Kleiber, M. (1972) Joules vs. calories in nutrition. *J. Nutr.* **102**, 309–312.

Kleiber, M. (1975) *The Fire of Life* (second edition). Robert E. Krieger Publ. Co., New York.

Kozlovsky, D. G. (1968) A critical evaluation of the trophic level concept. I. Ecological efficiencies. *Ecology* **49**, 48–60.

Krebs, C. J. (1978) *Ecology. The Experimental Analysis of Distribution and Abundance* (second edition). Harper and Row, New York.

Kucera, C. L. (1978) *The Challenge of Ecology* (second edition). C. V. Mosby Co., St Louis.

Leach, G. (1976) *Energy and Food Production*. IPC Science and Technology Press Ltd., Guildford, England.

Lehninger, A. L. (1971) *Bioenergetics* (second edition). W. A. Benjamin Inc., California.

Llewellyn, M. (1972) The effects of the lime aphid, *Eucallipterus tiliae* L. (Aphididae) on the growth of the lime *Tilia* × *vulgaris* Hayne. *J. appl. Ecol.* **9**, 261–282.

Llewellyn, M. and Qureshi, L. (1978) The energetics and growth efficiency of *Aphis fabae* Scop. reared on different parts of the broad bean (*Vicia faba*). *Ent. exp. appl.* **23**, 26–39.

Lindeman, R. L. (1942) The trophic-dynamic aspect of ecology. *Ecology* **23**, 399–418.

Lowe, G. D. (1978) The measurement by direct calorimetry of the energy lost as heat by a polychaete, *Neanthes* (= *Nereis*) *virens* (Sars). Ph.D. thesis. University of London.

MacArthur, R. H. and Wilson, E. O. (1967) *The Theory of Island Biogeography*. Princeton University Press, Princeton, New Jersey.

McNaughton, S. J. and Wolf, L. L. (1973) *General Ecology*. Holt, Rinehart and Winston Inc., New York.

McNeill, S. (1971) The energetics of a population of *Leptopterna dolabrata* (Heteroptera: Miridae). *J. Anim. Ecol.* **40**, 127–140.

McNeill, S. and Lawton, J. H. (1970) Annual production and respiration in animal populations. *Nature* **225**, 472–474.

Miller, R. J. and Mann, K. H. (1973) Ecological energetics of the sea weed zone in a marine bay on the Atlantic coast of Canada. III. Energy transformations by sea urchins. *Mar. Biol.* **18**, 99–114.

Miller, R. J., Mann, K. H. and Scarratt, D. J. (1971) Production potential of a seaweed-lobster community in eastern Canada. *J. Fish. Res. Bd. Can.* **28**, 1733–1738.

Needham, A. E. (1964) *The Growth Process in Animals*. Pitman and Sons, London.

Odum, E. P. (1971) *Fundamentals of Ecology* (third edition). W. B. Saunders Co., Philadelphia.

Odum, H. T. (1957) Trophic structure and productivity of Silver Springs, Florida. *Ecol. Monogr.* **27**, 55–112.

Odum, H. T. (1971) *Environment, Power and Society*. Wiley-Interscience, New York.

Open University, S 323 (1974) *Energy Flow through Ecosystems. Producers and Consumers*. The Open University Press, Milton Keynes, England.

Pamatmat, M. M. (1979) Anaerobic heat production of bivalves (*Polymesoda caroliniana* and *Modiolus demissus*) in relation to temperature, body size and duration of anoxia. *Mar. Biol.* **53**, 223–229.

Petrusewicz, K. and Macfadyen, A. (1970) *Productivity of Terrestrial Animals: Principles and Methods*. I.B.P. Handbook no. 13, Blackwell Scientific Publications, Oxford.

Phillipson, J. (1964) A miniature bomb calorimeter for small biological samples. *Oikos* **15**, 130–139.

Phillipson, J. (1966) *Ecological Energetics*. Edward Arnold, London.

Pimentel, D., Dritschilo, W., Krummel, J. and Kutzman, J. (1975) Energy and land constraints in food-protein production. *Science* **190**, 754–761.

Pimentel, D. and Pimentel, M. (1979) *Food, Energy and Society*. Edward Arnold, London.

Potts, W. T. W. (1954) The energetics of osmotic regulation in brackish- and fresh-water animals. *J. exp. Biol.* **31**, 618–630.

Ramsay, J. A. (1971) *A Guide to Thermodynamics*. Chapman and Hall, London.

Ricker, W. E. (1946) Production and utilization of fish populations. *Ecol. Monogr.* **16**, 375–389.

Ricklefs, R. E. (1980) *Ecology* (second edition). Nelson and Sons Ltd., Sunbury-on-Thames, England.

Rochereau, S. and Pimentel, D. (1978) Energy tradeoffs between Northeast fishery production and coastal power reactors. *J. Energy* **3**, 545–589.

Russell-Hunter, W. D. (1970) *Aquatic Productivity*. Macmillan, London.

Schneider, S. H. (1976) *The Genesis Strategy. Climate and Global Survival*. Plenum Press, New York.

Schroeder, L. A. (1980) Consumer growth efficiencies; their limits and relationships to ecological energetics. Unpublished manuscript.

Slobodkin, L. B. (1959) Energetics in *Daphnia pulex* populations. *Ecology* **40**, 232–243.

Slobodkin, L. B. (1960) Ecological energy relationships at the population level. *Am. Naturalist* **94**, 213–235.

Slobodkin, L. B. (1962) "Energy in animal ecology", in *Advances in Ecological Research*, Vol. 1, ed. J. B. Cragg, Academic Press, London, pp. 69–101.

Slobodkin, L. B. (1964) Experimental populations of hydrida. *J. Anim. Ecol.* **33** (suppl.), 131–148.

Slobodkin, L. B. (1970) "Summary", in *Marine Food Chains*, ed. J. H. Steele, Oliver and Boyd, Edinburgh, pp. 537–540.

Slobodkin, L. B. (1972) "On the inconstancy of ecological efficiency and the form of ecological theories", in *Growth by Intussusception* (Ecological Essays in Honor of G. Evelyn Hutchinson). *Trans. Conn. Acad. Arts Sci.* **44**, pp. 293–305.

Solomon, D. J. and Brafield, A. E. (1972) The energetics of feeding, metabolism and growth of perch (*Perca fluviatilis* L.). *J. Anim. Ecol.* **41**, 699–718.

Somero, G. N. and Hochachka, P. W. (1976) "Biochemical adaptations to temperature", in *Adaptation to Environment: Essays on the Physiology of Marine Animals*, ed. R. C. Newell, Butterworths, London–Boston, pp. 125–190.

Spanner, D. C. (1964) *Introduction to Thermodynamics*. Academic Press, London and New York.

Stearns, S. C. (1977) The evolution of life history traits: a critique of the theory and a review of the data. *Annual Review of Ecology and Systematics*, Vol. 8, ed. A. Macfadyen, Academic Press, London.

Stewart, M. G. (1979) Absorption of dissolved organic nutrients by marine invertebrates. *Oceanogr. mar. Biol. ann. Rev.* **17**, 163–192.

Stirling, J. L. and Stock, M. J. (1973) "Non-conservative mechanisms of energy metabolism in thermogenesis", in *Energy Balance in Man*, ed. M. Apfelbaum, Masson, Paris, pp. 219–227.

Taylor, L. R. (1975) Longevity, fecundity and size; control of reproductive potential in a polymorphic migrant, *Aphis fabae* Scop. *J. Anim. Ecol.* **44**, 135–163.

Teal, J. M. (1957) Community metabolism in a temperate cold spring. *Ecol. Monogr.* **23**, 41–78.

Teal, J. M. (1962) Energy flow in the salt marsh ecosystem of Georgia. *Ecology* **43**, 614–624.

Thomas, S. P. and Suthers, R. A. (1972) The physiology and energetics of bat flight. *J. exp. Biol.* **57**, 317–335.

Tilly, L. J. (1968) The structure and dynamics of Cone Spring. *Ecol. Monogr.* **38**, 169–197.

Tucker, V. A. (1975*a*) "Flight energetics", in *Avian Physiology*, ed. M. Peaker, *Symp. zool. Soc. Lond.* **35**, pp. 49–63.

Tucker, V. A. (1975*b*) The energetic cost of moving about. *Am. Scient.* **63**, 413–419.

Weir, J. B. de V. (1949) New methods for calculating metabolic rate with special reference to protein metabolism. *J. Physiol.* **109**, 1–9.

Welch, H. E. (1968) Relationships between assimilation efficiencies and growth efficiencies for aquatic consumers. *Ecology* **49**, 755–759.

White, A., Handler, P., Smith, E. L., Hill, R. L. and Lehman, I. R. (1978) *Principles of Biochemistry* (sixth edition). McGraw-Hill Kogakusha Ltd., Tokyo.

Whittaker, R. H. (1975) *Communities and Ecosystems* (second edition). Macmillan Publishing Co. Inc., New York.

Wiegert, R. G. (1964) Population energetics of meadow spittlebugs (*Philaenus spumarius* L.) as affected by migration and habitat. *Ecol. Monogr.* **34**, 217–241.

Wiegert, R. G. (1968) Thermodynamic considerations in animal nutrition. *Am. Zoologist* **8**, 71–81.

Wilkie, D. R. (1976) *Muscle* (second edition). Edward Arnold, London.

Woledge, R. C. (1980) "The heat production of intact organs and tissues", in *Biological Microcalorimetry*, ed. A. E. Beezer, Academic Press, London, New York, pp. 145–162.

Index